# PROBLEMATIC REALITY

The hyper-inflated marketing of climate change

John Matic

ISBN 978-0-646-55277-4

The author can be contacted at:

matic.john@gmail.com

www.problematicreality.com

# PROBLEMATIC REALITY

If global warming is *unequivocal*, the word used by the UN's Inter-governmental Panel on Climate Change to promote the most vital issue facing our planet, then shouldn't the detailed temperature records, measured each day around the globe, verify this claim?

Take a journey through the climate veneer into a world of bad statistics and oversimplifications. Not because of incompetent scientists, although there are a few, but because the climate agenda is utilising the techniques of *hyper-inflated marketing,* a powerful tool usually reserved to promote nationalism and military conflict.

*Problematic Reality* will change the way you view the world and the way to solve its problems.

# AUTHOR

John Matic lives on a family farm in Appin, Australia with his beautiful wife, two sporty kids, a collection of farm animals, and a never ending list of projects to finish.

To Jenelle,

As you tell me, life's a marathon not a sprint. Thank you from the
bottom of my heart.

Love, John

# TABLE OF CONTENTS

# Chapter 1
# A 1000 YEAR DROUGHT

Did global warming cause the drought on our family farm? It seemed like such a simple question.

You see, we live in Appin, a small country town located about 70 kilometres southwest of Sydney. Appin is not what you'd call a prominent place, but it was brought to national attention in 1979, when 16 miners died following an underground explosion in the town's coal mine. While coal mining is the major employer in Appin, its proximity to Sydney and Wollongong allows a major part of the community to commute for employment. We fall into this category. No one in our family is associated with the coal industry. In fact, BHP plans to extend their mine under our property in the next few years, so I'd have to say that I am not a fan of coal mining.

I was born and raised in Pittsburgh, Pennsylvania. After graduating in Computer Science from Penn State, I went to work for IBM in upstate New York. From here I landed an international assignment to Sydney, Australia. I ended up staying. That was in October 1985. My beautiful wife is from Appin. So here we are.

Farming was vital to the town over the past 200 years. But while the land still looks rural, the major agricultural enterprises that once dotted the landscape have somewhat faded away. Don't get me wrong, there are still cows in the fields and the Ingham's chicken facility continues to ship hundreds of chooks past our door each day. We like to tell the kids that they are going on a holiday to Kentucky. We still have stock sale yards in nearby Camden and our land is zoned rural - *for agricultural use only* – but the urban expansion of Sydney is getting closer each year.

Our farm is 33 acres in the old scale – 12 hectares in the new; running a small herd of cattle and alpacas. We also have some 30 chickens. Every year I plan to get turkeys. My daughter Hannah dreams of being an animal doctor, so we keep collecting various types of animals from time to time.

My wife's father ran a piggery on the property next door in the 1960s. We were lucky to purchase our land. Basically, we bought back the family farm. Yet, the stark reality of living on the land hit home on Christmas Day in 2001, when a large bush fire almost burned our new home to the ground. We had only just a week prior exchanged contracts. Our good fortune, since I don't think we had any insurance to cover a bush fire.

I was also lucky to have my father-in-law's experience about the property and the region. We were hit by a major drought that started just after we moved onto the property. By 2004, as the drought continued and the land became parched, I started to question why anyone would want to be a farmer. Having to purchase expensive hay for our newly established animals clearly made me understand that the climate is what made farming either work or fail. And with the price of fodder more expensive than the cattle's purchase price, the financial impact of the drought became apparently clear.

The venomous snakes came to the house looking for water as well. A brown snake killed our dog Steeler – a black and gold kelpie. That was a very sad day.

Over the same period of time, reports of global warming and climate change became more common. The general term *environment*, used extensively during the 1990s, became replaced by climate terms like carbon footprint and the greenhouse effect.

As the media coverage of global warming intensified, like many people, I became certain that the drought must have been caused by climate change.

On our farm we have a rain gage and a thermometer but our historical records were limited to our recent purchase date. I was determined to figure out just how much our temperatures have increased over time - this would prove our drought was caused by climate change.

My wife would say I was obsessed.

Now Appin was one of the earliest settlements in Australia, founded in 1810. Hamilton Hume and Captain William Hovell set off on their *Voyage of Discovery* – the Australian equivalent to the USA's Lewis and Clark - from Appin in 1824.

It seemed sensible to believe that detailed weather records should exist for such an historic site.

And of course weather records should be a certainty because Appin is home to the Cataract Dam, built between 1902 and 1907. It was the first dam built in the upper Nepean scheme of Sydney's water catchment system. This project was a significant feat of engineering for its day and involved considerable planning. One would think that measuring the rainfall and the weather in the dam's catchment would be a prerequisite to building the structure?

Maybe not.

It seems that politics existed even 100 years ago. So much so that the famous Australian bush poet, Banjo Paterson, wrote the satirical ballad - *The Dam that Keele Built* - to immortalise the controversy of the day. Its final verse goes as follows:

> This is the Sydneyite afraid
> That a serious blunder will be made
> Because of the Minister, quite dismayed
> At the sight of the Scornful Mr Wade,
> Who sneered at the Calculations made
> By the Eminent Engineer by Trade,
> Head of the Water and Sewer Brigade,
> Who measured the steam that brought the water to fill the dam that Keele built.

So the only thing that has changed in a century is we are now Sydneysiders rather than Sydneyites.

But get this. Forget about historical settlement or a 100 year old dam to suggest the holding of weather records. The Australian Bureau of Metrology, affectionately known as the BOM, has its weather radar station for Sydney located in – you guessed it - Appin.

At this point it hit me; it must be possible, maybe even easy, to find historical weather information for Appin on the internet. So with purpose, I began my climate journey, starting with the historical climate information on the BOM's website.[1]

However, this is where my story takes its first turn.

While Appin has weather records in the BOM database, they only recorded rainfall rather than the temperature data I was initially seeking. But rainfall was a good start. We were trying to understand the drought after all.

The rainfall records started in 1904, two years after the dam's construction began. As I said, you would think they would have measured rainfall for years prior to building such a project. If they did, this data is missing.

During construction in 1904, the dam's catchment received a massive 1448mm of rainfall during the year. But can you imagine the fallout after building the dam and failing to get adequate rainfall to fill it up?

By 1909, only 635mm of rain fell in the year. By the end of the First World War, Cataract Dam was yet to receive more monthly rain then it had in 1904. Refer to the graph of monthly precipitation in Figure 1.

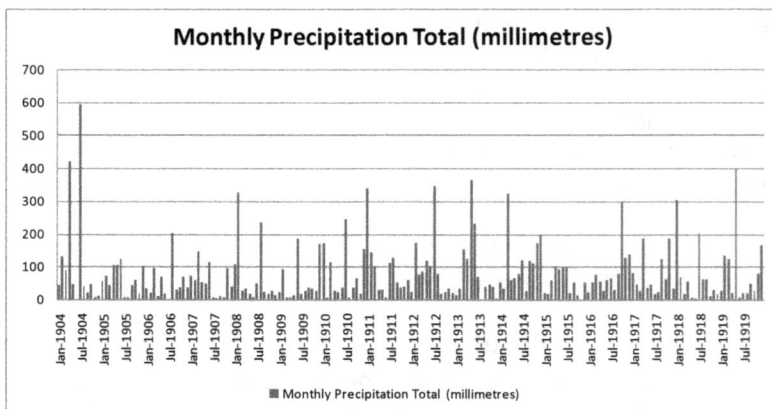

Figure 1 – Monthly rainfall at the Cataract Dam during the early years from 1904–19. In 1904, 1448mm of rain fell on the catchment during the year, with 420mm falling in the month of April and 594mm in July. Yet in 1908, the year they needed the water to fill the newly constructed dam, only 635mm of rain fell with the largest single month being only 185mm in June.

It seems Appin fits the description well of the famous Australian poem, *My Country*. First published in 1908 by Dorothea Mackellar, its second stanza goes as follows:

> *I love a sunburnt country*
> *a land of sweeping plains*
> *of rugged mountain ranges*
> *of drought and flooding rains*

If Dorothea knew in 1908 that the flooding rains come after a drought, why are we always surprised when this happens? During the 106 years of records, Appin has had its fair share of drought and flooding rains.

The historical rainfall data suggested that we could expect about a meter of rain each year. Certainly there have been many years below this level. The data also showed a long drought from about 1935 to 1945. In 1950, a record deluge of 2293mm of rain fell. 1968 was a very dry year. This was when my wife's father sold his piggery. He often talked about how dry it was when he worked the land. One day, we went for a walk down to the Georges River at the back of our property. He looked on with pride when we found the remains of the pump house he built over 50 years ago to supply his family with water.

While the historical record went back to 1904, what attracted my attention was a small button on the BOM webpage, just below the nearest bureau stations. You could tick *"only interested in open stations."*

I was stunned by how many weather stations were closed. Stanhope 1.1km away – closed, Cataract River 3.8km away – closed, Wedderburn 5.8km away – closed, Wilton Post Office 8.5km away – closed, Broughton Village 12.3km away – closed. At least Cataract and Appin were still operational.

The other complication was that of missing data. This is a common theme repeated throughout my analysis - and a very important one. From 1904 to 1998, a full 12 months of data records are contained in the database for each year.

In 1999, only 10 months is available. More recently, 2005 and 2006 contain 11 months, 2007 has 7 months, 2008 only 3 months and 2009 has 10 months. Data for 2010, when it seems like it rained everyday and the dam on our farm is busting over, there is only 1 month recorded in the database. How are we able to measure every day from 1904 to 1998, 94 years of data, but fail to collect this data when the focus on climate change is so great?

In addition to the problems of missing data, linear trend lines have got to be one of the most misused tools in statistics. To explain this, a dataset is simply a collection of records. As such, the precipitation values, shown in the graph, are the dataset. When plotted, if this data looks like a straight line, the dataset is called linear. The word *linear* actually means *resembling a straight line*. Now look at the rainfall data for Cataract Dam in Figure 2, does it look like a straight line? Of course it doesn't, the data is all over the place.

Then to make matters worse, in other words, less statistically valid, the raw data is often averaged and missing data is ignored. So the daily rainfall measurements of say 27 days, with 3 days missing, are converted to a monthly average, which is then converted to an annual average. This is often done to make the raw data look less confusing, forgetting that the confusing nature of the data is in itself a very useful clue. The more you average something, the more it looks like a straight line. But you know from the raw data that it isn't a straight line.

Now comes the big mistake. The highly averaged data, typically given an easy to understand name, in this case *mean annual rainfall*, uses the linear trend line to make a conclusion about future rainfall. So back to the Cataract Dam data in Figure 3. Taking the starting point of the trend line as 995mm and the ending point as 1250mm, the claim could be made that *rainfall at the dam has increased 25.6% since it opened for operation in 1904.*

With the pressures in society and the media to simplify messages, data is often presented as a linear trend line of the mean and future trends are forecast with just one number.

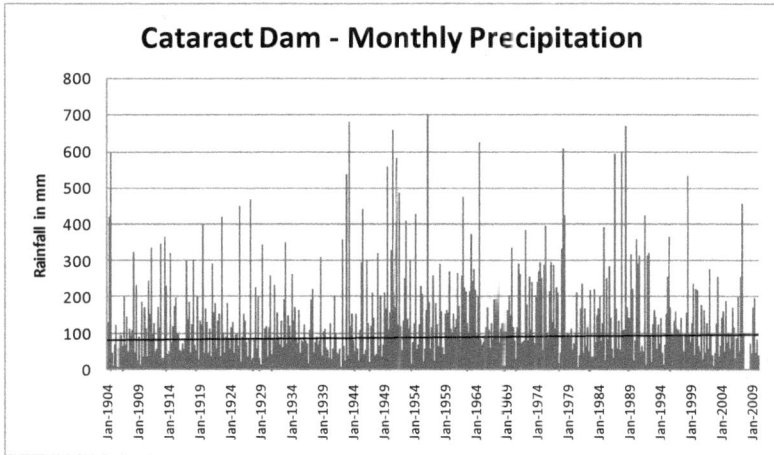

Figure 2 - Monthly rainfall for Cataract Dam. This data shows that our drought from 2001-07 was not unusual and certainly far worse droughts have occurred over the past 106 years. The linear trend line shows a slight increase in rainfall over the years.

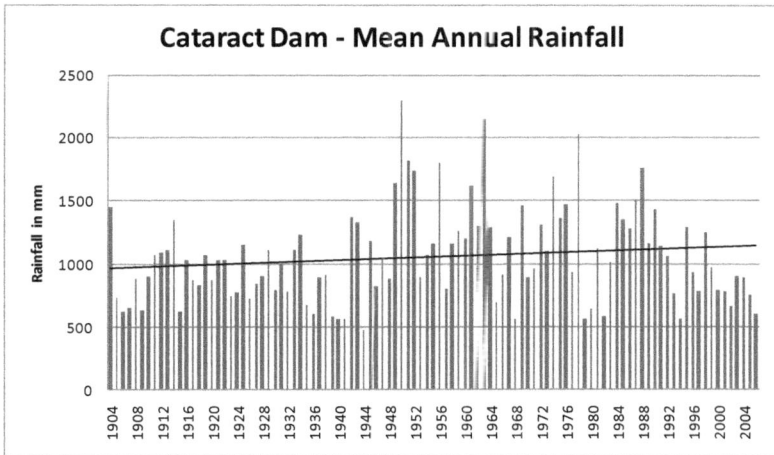

Figure 3 – Mean annual rainfall for Cataract Dam. The drought from 2001-07 was widely reported in the media as the worst in 1000 years and the result of climate change. This rainfall chart suggests much drier periods over the past 106 years. Also, notice the difference between the trend line of the annual and monthly totals.

A more accurate way to analyse this data is to place each of the raw records into a group that is representative of the range of the data. In statistics, this is called a bin. So for rainfall, the bin might be how many days when there was no rainfall, less than 5mm, 5-10mm and so on. This analysis produces a distribution of the rainfall. Often this is called the *bell curve*, but a bell shape is just one of the thousands of ways the data can appear.

So for Appin's Cataract Dam, the daily rainfall for the years from 1904-1990 are compared to 1991-2009. This tells us that there was a 69.1% chance of no rainfall on any given day and now there is a 62% chance of no rainfall; so we now have more rainy days. However, the only significant difference in rainfall between the two distributions is an 8% increase in days where 5mm of rain fell - a far cry from the linear forecast of a 25.6% increase in rainfall.

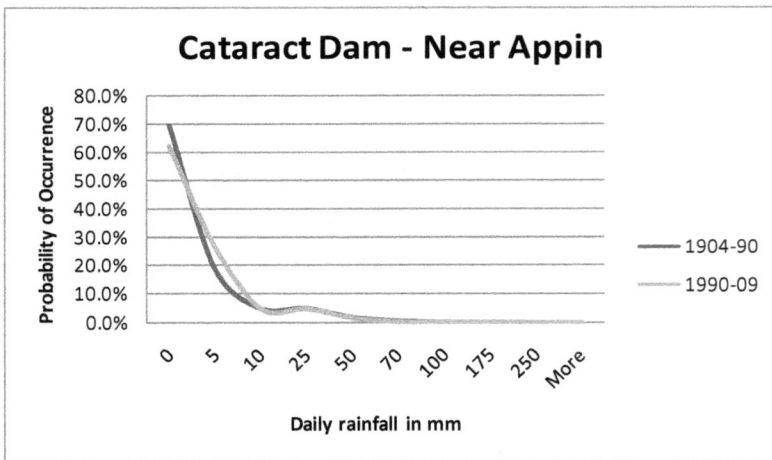

**Figure 4 - Rainfall distribution for Appin's Cataract Dam actually shows more rainy days with an increase in days with 5mm of rain and a very slight decrease of days with 10-25mm of rain.**

So armed with 106 years of data, the media pronounces our current drought is the worst in 1000 years. The scientists claim it is caused by changing rainfall patterns from global warming.

This story from the science magazine Cosmos, in December 2006, paints the picture of the general media landscape very well:[2]

SYDNEY: Australia's current drought, called the worst in 1,000 years, is the result of changing rainfall patterns and may necessitate major changes in the continent's water economy.

Experts cited climate change as a factor contributing to the increasing uncertainty in Australian weather.

"It's a combination of short El Nino drought and longer-term decreasing rainfall," said Michael Coughlan, of the Australian Bureau of Meteorology. "The combination of short and long-term drought is surprising - we didn't see it coming, and it's really shaken everyone up."

The findings come as part of the World Meteorological Organisation (WMO) report on the 2006 global climate, made public today. 2006 was the 6th hottest year on record globally, according to the U.N.'s weather service, and saw prolonged droughts in Australia, the U.S., Brazil, and the Horn of Africa.

Some Australian experts don't see rainfall on the arid continent increasing again anytime soon.

"Drought is too comfortable a word," said John Williams, the New South Wales state Commissioner for Natural Resources. "Drought connotes a return to normal. We need to be adjusting."

According to Williams, Australia is a nation of extremes, where droughts and flooding rains are the norm. The last 50 to 60 years, when Australia developed much its water infrastructure, have been times of relative plenty, he said.

He harked back to the years between 1900 and 1950, when rivers in the Murray-Darling system were dry for a total of 17 years. "It's only been dry 5 years since then," he said. According to Williams, the continent is reverting to the drier conditions of the past, exacerbated by climate-change induced uncertainty.

It's a return to the sequence of the first 50 years [of the century]," he said.

Williams and others think that Australia's new rainfall pattern will require fundamental changes in the way water is used.

The years of high rainfall have led to an over-allocation of water resources that we can no longer sustain, said Williams. "We need more water storage and desalinisation, or better water usage."

Jenifer Simpson, an industrial chemist and water advocate, agreed, stressing water recycling as the way to reduce reliance on uncertain rains. "Right now our 'water cycle' is not a cycle," she said. "Our current urban water cycle is a straight line from dam to disposal, with a shortage of water at one end and pollution at the other."

She said that the technology exists to make recycled water safe for drinking, but that a lack of understanding between the water industry and the community prevented recycled water's acceptance.

"Recycling should be accepted and exploited," she said.

- 	Cosmos Magazine, Benjamin Lester, 15 December 2006

Banjo Paterson would have been proud - *It seems the minister is quite dismayed.* The four major dams of the Upper Nepean water catchment were built between 1904 and 1936. Sydney's Prospect Reservoir was built in 1888. Even the Sydney Catchment Authority's website details how the Warragamba Dam, Sydney largest dam, was first identified as a site in 1845, but it was not until the "record drought" of 1934 to 1942 that caused the dam's construction to start in 1945. It was finished in 1960.

So how can the NSW Commissioner for Natural Resources make the claim that the water resources were built in times of relative plenty? The distribution of rainfall is virtually unchanged since 1904.

Was the current drought a product of climate change? Can we expect a new rainfall pattern that will result in prolonged drought? Or is it just an ordinary drought?

Well the NSW Government went for the prolonged drought option. On 28 January, 2010, it opened its newly constructed water desalination plant in the Sydney suburb of Kurnell. The Government spent $1.8 billion to produce an estimated output of 250 million litres of water per day. This is approximately 15% of Sydney's water supply.[3]

Yet the capital cost does not include the electricity bill for running the desalination. Australia's peak scientific body, the CSIRO, calculated that these plants use seven times more electricity than conventional water treatment plants. What this simply means is higher water bills. But don't worry, we are told that this is good, since higher water bills will give consumers more incentive to use water wisely.[4]

Sydney is not alone in its desalination effort. Adelaide and the Gold Coast all have spent billions on these facilities. Perth is building its second plant. Water bills in Melbourne will increase by nearly one third once the nation's largest plant, at $3.5 billion comes on line in late 2011.

The flooding rains fell in 2010 and continued into 2011. No need to drink recycled sea water just yet. Again, we are always surprised by the flooding rains. Perhaps we get a bit too sunburnt during the droughts.

So after the *worst drought in 1000 years*, the Sydney Catchment Authority reports dam levels to be at 72.5% as at 3:00pm on 30 December 2010.[5]

Our farm is very wet as well. Our water tanks and farm dam are all full. Yet the Bureau of Meteorology acknowledges that while we are experiencing our wettest year since 2000, senior climatologist, Dr Blair Trewin, says it's too early to declare Australia's decade-long drought is over.[6]

**Figure 5 - The rain water is collected from the roof and flows into the rain tank. In this photo, taken during a thunderstorm, we have too much water and our tank is overflowing. Living so close to Sydney's water supply, we are actually not connected to the mains and we have to fend for ourselves. As such, you can only take a long shower while it is raining. However, during each rainstorm, I first need to climb a wet ladder to clean the gutters. Visitors are always surprised that no further purification is performed; we drink the water straight from the roof. Photo: Jenelle Matic**

# Chapter 2
## CLIMATE CHANGE

Global warming is presented as very simple model. It is widely documented in the scientific, general media, government and academic community. An entire division of the United Nations, the Intergovernmental Panel on Climate Change (IPCC), has been jointly established by the United Nations Environment Programme (UNEP) and the World Meteorological Organization (WMO). (The IPCC is referenced often in this book, so it is an important acronym to remember.) This group produced the main reference report on global warming, called AR4; for the Fourth Assessment Report: Climate Change 2007.[7] It warns that climate change is *"unequivocal"* and evident from observations of increases in global average air temperatures.

The IPCC model for climate change goes as follows: Human pollution, mainly from burning coal, has been placing increasing levels of $CO_2$ in the atmosphere. This gas causes the air temperature to increase from a greenhouse effect, leading to a warmer planet. This warming leads to a wetter world, as it is argued that moisture increases with temperature. The increased moisture is in itself also a greenhouse gas and the temperature continues to increase, producing a changed climate that will have drastic effects on our planet and way of life.

In this new climate, we will have less cold days, more hot days and more record hot days. The IPCC Model for this is shown in Figure 6. This graphic was taken directly from the AR4 document and it was determined to be so important that it is copyrighted to the IPCC.[8]

The IPCC goes on to say that this increase in mean temperature will lead to weather instability and more extreme weather events such as cyclones, typhoons and hurricanes. The warmer temperatures will cause glaciers and ice sheets to melt, resulting in rises to sea levels, engulfing small island nations and causing severe coastal erosion.

In its book *Climate Change, what you can do about it*, the CSIRO states that "climate change is the most vital issue on the planet."[9] This is a view that has been echoed by many Governments and the media around the globe. The focus on global warming is so significant that even the Nobel Peace Prize in 2007 was awarded jointly to the Intergovernmental Panel on Climate Change and Al Gore "for their efforts to build up and disseminate greater knowledge about man-made climate change, and to lay the foundations for the measures that are needed to counteract such change."

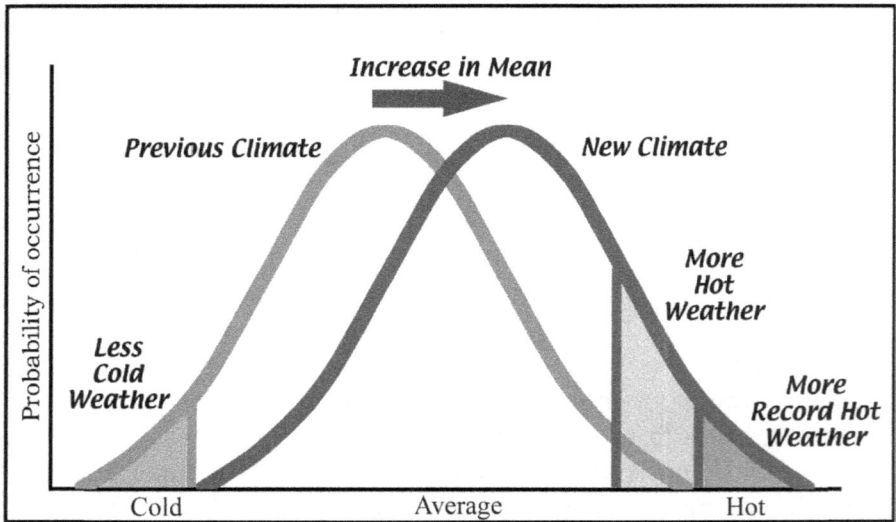

**Figure 6 - IPCC AR4 Model for climate change showing an increase in the mean temperature leads to a new climate with less cold weather, more hot weather and more record hot weather. ©IPCC 2007.**

The observations that lead to these conclusions include the AR4 claim that "eleven of the last twelve years (1995-2006) rank among the twelve warmest years in the instrumental record of global surface temperature since 1850."

Obviously to make such a claim the world's climate scientists must have a complete record of data that categorically shows these global warming trends?

It is also logical to assume that the answer regarding the drought on our farm and whether it has been caused by climate change lies in finding these historical temperatures to see how much our climate has warmed. It really disappointed me that the historical temperature data for Appin did not exist. Surely if they measured the rainfall they would also take the temperature? I couldn't figure that out.

After reading scores of scientific journals and extensively searching the web, it appeared to me that the starting point for the observed global temperature increases was not in Australia but rather on the Antarctic Peninsula, from work conducted by the British Antarctic Survey (BAS). Their findings, *Devil in the Detail,* initially published in the journal *Science,* 7 September 2001, claimed that warming in the Antarctic Peninsula was occurring faster than anywhere else on the planet – "warming at low elevations on the western coast of the Antarctic Peninsula is as large as any increase observed on Earth over the last 50 years."[10]

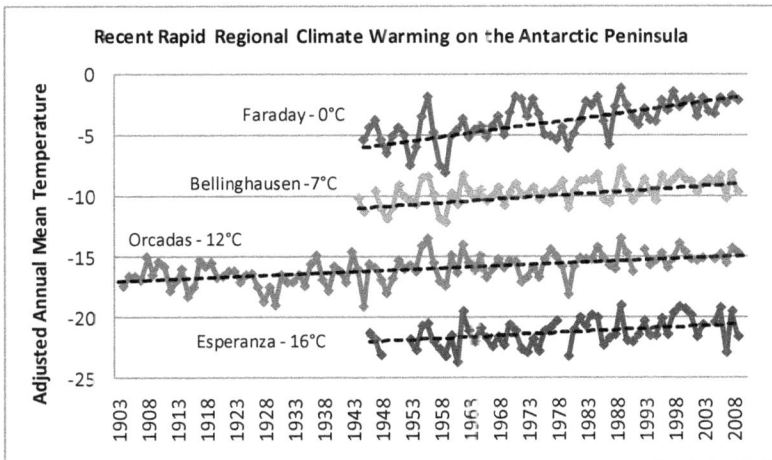

**Figure 7 – British Antarctic Survey chart showing the warming claims on the Antarctic Peninsula. The average annual temperature for each of the locations is reduced by a fixed temperature, as shown, to fit the data on one chart.**

In 2003, they also tried to coin a phrase, Recent Rapid Regional Climate Warming (RRR), to describe the localised warming of regional zones.[11] It didn't catch on. The BAS followed up these papers publishing works covering the 50 years of Antarctic climate change in 2005[12] and the melting conditions on the peninsula in 2006[13]. A web site containing their temperature database is called READER - REference Antarctic Data for Environmental Research.[14]

It contains mean monthly and mean annual temperatures for many locations. The data was plotted by the BAS with linear trend lines added, forecasting rapid temperature increases. The four locations of Esperanza, Orcadas, Bellinghausen and Faraday formed the core proof by the BAS that temperatures on the Peninsula are dramatically increasing. Refer to Figure 7.

The initial BAS publications made Antarctica the focus for global warming. Kind of a poster child of melting ice.

So you had it. The Antarctic Peninsula is dramatically warming. The ice is melting. Sea levels will rise. Our farm is in drought. Global warming is upon us. The planet is threatened.

Could it really be that easy?

# Chapter 3
## MY SISTER KATE

I like to call her a university professor, but actually she teaches in a community college. And since Kate did some work for NASA, I also like to think of her as my rocket scientist sister. She holds her Bachelor's Degree in Physics and a Masters in Statistics.

She is also a total greenie.

She lives in Flagstaff Arizona, the self proclaimed capital of the green movement. She refuses to trim a tree branch in her back yard. She took up cross country skiing to save the environment. She totally believes in climate change.

What troubled me when I read the BAS papers and examined the READER database is that all the data which formed their conclusions were based on linear trend lines of annual averages. Could you draw such conclusions on global warming based on annual or even monthly averages? Also, since temperatures can change by the minute, what daily temperature do you average? After all, there is a high temperature, a low temperature and some type of average temperature for each day.

Another good question, how do you account for missing or unknown temperature records? If on some freezing, dark winter's day in the middle of nowhere in Antarctica, the lonely meteorologist did not make a recording, did this really matter? What if they missed 20 days in a year or 200?

So I sent my sister an email asking her to analyse some data. I didn't tell her that it was temperature data. I just gave her the data and the graph based on Esperanza, with the trend line and asked her if she could draw the conclusion that the data was increasing over the samples.

Her response was interesting. She wrote:

"I personally don't see a connection between the averages increasing over the samples. Also, if you want to show something is increasing, it isn't really the best to use averages. It is better to use raw data and do a paired sample comparison. However, you can't do that because of the missing data. You can do a test that compares the means, but that is for independent data, one where the values have no correspondence to each other. In this case, each sample seems to be taken using the same time interval. That makes me think they are not independent.

As for using the means and then doing a simple trend line, it is not the best idea. You lose a lot of information using the means. You see this when they talk about the stock market. They like to say that it is going up or down by looking at the means, but in reality they are very volatile. Also, looking at the graph you created, given that the y-axis didn't start at 0, it makes the increase seem more dramatic. But in reality it is very slight increase or not at all."

Wow!

Did my sister really just punch a hole in the work done by the British Antarctic Survey and the UN Intergovernmental Panel on Climate Change?

Remember the monthly rainfall for Appin in Figure 2? Its linear trend line is nearly flat, implying consistent rainfall yet the annual rainfall chart, Figure 3, has the linear trend line increasing, implying an increase in rainfall. The same data, shown monthly and annually, gives two different results. Yet neither of these trend lines is correct, as we are in a 1000 year drought!

Following my sister's advice, I was determined to see if the raw temperature data would show the same conclusions.

I started reading climate paper after climate paper. They all used the same metric – mean annual temperatures. I could not find one of them that used raw temperature data. Even the IPCC's AR4 used annual means.

What did the raw temperature data look like?

After searching, I found a website run by the United States Department of Commerce's National Climate Data Centre. Here they had the Global Summary of the Day (GSOD) which contained daily temperatures from all around the world.[15] I selected Antarctica and the base of Esperanza, used by the BAS in support of their warming conclusions.

The database was very interesting. It contained a wide variety of weather data including three temperatures for each day, the maximum, minimum and mean.

Additional research determined that the mean temperature is calculated in a different manner all over the globe. It could be the difference between the maximum and minimum; or it could be based on the average of the temperature taken each hour; or a sample average taken a number of times per day. I don't want to dive too deep here, but the important thing is that the daily mean temperature is not recorded in the same manner around the world or historically. This makes comparison of this mean virtually impossible, yet it is used widely by the IPCC.

So with the GSOD data, I decided to focus on the maximum and minimum raw daily temperatures. After all, the IPCC model in Figure 6 claims that we should experience less cold days, more hot days and more record hot days. The minimum would tell us if we are experiencing less cold days and the maximum would show us if we are experiencing more hot days.

GSOD data for Esperanza existed from 1973 onwards. I simply ran the report on the internet, downloaded the data, put it into a spreadsheet and graphed the maximum and minimum temperatures. The first thing I noticed in the graph was a massive range of data, the difference between the low and the high. The daily maximums were as low as -28°C. This temperature occurred on 26 July 1994. The warmest maximum, a *'balmy'* 22°C, occurred 4 times in 1973, 74, 75 and 77. These measurements however look out of place and you would have to question their validity. Either way, Esperanza would not be a very nice place to live.

Esperanza is an Argentine scientific base, famous for the birth of Emilio Marcos Palma, the first person to be born in Antarctica on 7 January 1978. Emilio enjoyed a maximum temperature of 6.7°C on his first day of life. Esperanza is located on Hope Bay, just at the very tip of the Antarctic Peninsula.

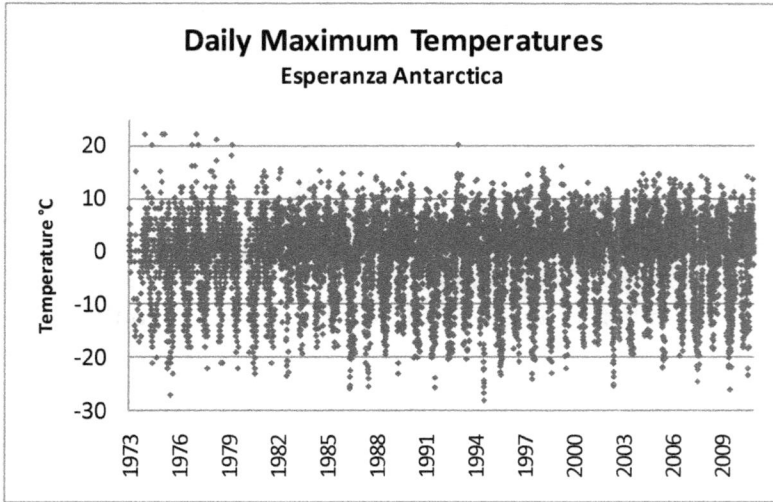

**Figure 8 - Daily maximum temperature data for Experanza Antarctica since 1973. It shows a very dynamic temperature range but no obvious signs of warming.**

Here we have 12,137 daily temperature records from 1973 to 2010. We are missing 1723 days over the period including a big chunk from mid-September 1979 to mid-February 1980. The sample gets better over time, as a total of 12.4% of the days are missing from 1973, yet only 5.7% of the days are missing from 1983.

I played with the data in Figure 8 for days. I couldn't get over the shock that the graph of 12,137 daily temperatures showed no obvious increase in temperatures from global warming. No more hot days and no fewer cold days. The climate looked unchanged in Esperanza since 1973.

Actually, I'm being kind. It sort of looked like the temperatures for Esperanza might be getting cooler. There were quite a few days in the 1970's over 20°C, but only one day since, 20°C in 1992.

Esperanza was one of the four locations the British Antarctic Survey used to prove global warming to the world.[16] The temperature increases form the foundation for the UN Intergovernmental Panel on Climate Change.

In analysing the BAS findings, the first step was to extract the READER graph for the individual site of Esperanza, shown in Figure 9. Then, following the same model used by the IPCC (Figure 6), I took the most complete data records (1983-2010), split these in half and determined the statistical distribution of both samples. This produced a graph showing very similar climates, shown in Figure 10.

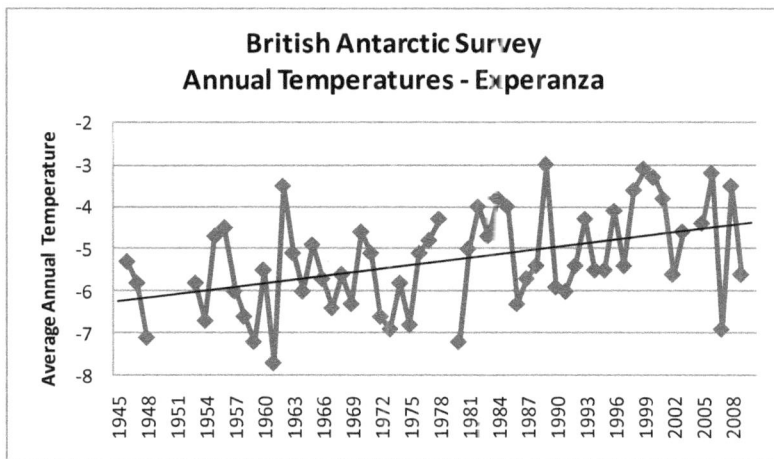

**Figure 9 - The British Antarctic Survey READER temperature records for Esperanza were used to prove the theory of global warming.**

I then took the mean annual temperatures from the British Antarctic Survey and plotted these on an axis representative of the actual temperature range, from -30°C to +25°C as shown in Figure 11.

The soaring temperatures shown by the BAS have vanished when used against a scale consistent with the range. Compare Figure 9 with Figure 11.

The BAS READER data is an annual average of the mean daily temperature, so to see a clear picture of the representative nature of this average, the daily minimum and maximum temperatures are overlaid with the annual average. This chart is shown in Figure 12.

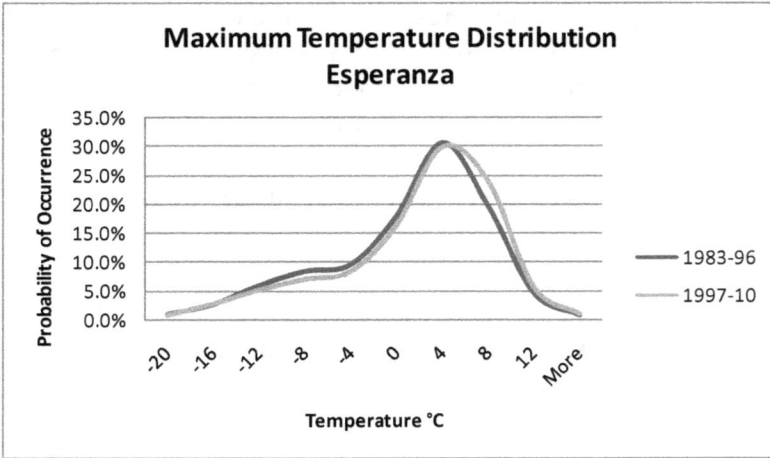

**Figure 10 - Maximum temperature distributions for Esperanza, comparing the years 1983-96 against 1997-10. The climates are very similar, with only slight differences between the two limited samples.**

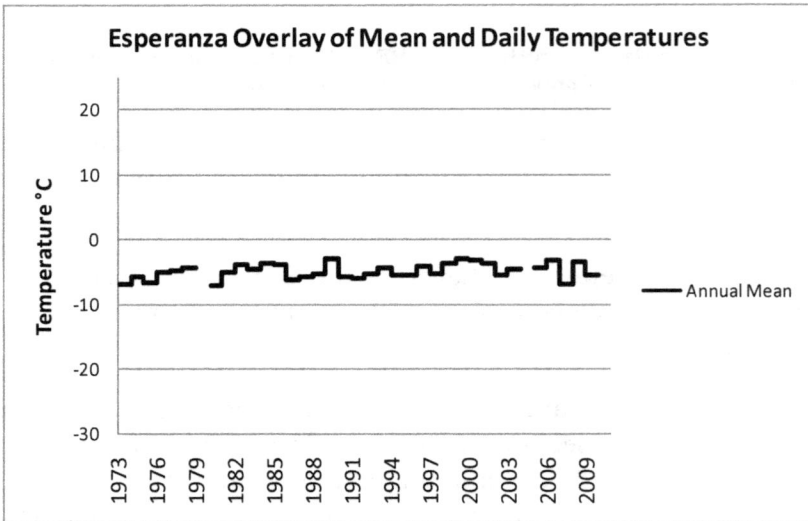

**Figure 11 - Mean annual temperatures for Esperanza plotted on a scale typical of the range of temperatures experienced at the Argentine base. Soaring temperatures are not longer evident.**

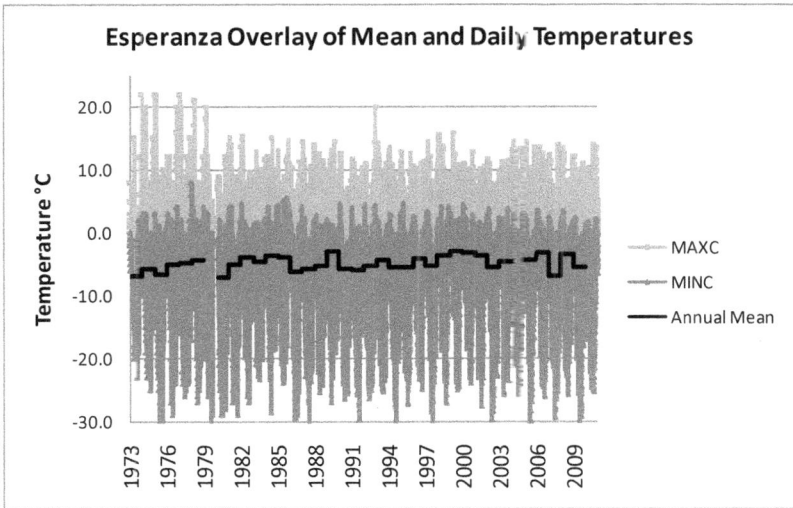

**Figure 12 - The daily minimum and maximum GSOD temperature with the BAS READER annual mean temperature overlaid.**

At this point I feel it is vital to re-stress what we are analysing. The British Antarctic Survey and the UN Intergovernmental Panel on Climate Change have identified that climate change on the Antarctic Peninsula was occurring faster than anywhere else on the planet.[17]

In other words, this is as bad as it gets. But there are no obvious signs of warming in Esperanza.

Was Kate right in the need to examine the raw data?

# Chapter 4
## ANTARCTICA

The bottom of our planet holds a world virtually untouched by human hand. While I've never been to Antarctica, in my previous job, I was lucky enough to help sponsor the restoration of Mawson's Huts.

In 1912, Australia's Sir Douglas Mawson led 18 brave souls in the Australasian Antarctic Expedition's journey to Cape Denison on Commonwealth Bay. As a geologist, he chose the site and built his huts in an area that on first inspection looked very calm. Little did he know that Cape Denison was perhaps the windiest place on the planet. Soon after arriving, the weather returned to its normal state of blizzard and his party was forced to endure severe hardship. They happily left in 1913. The site was revisited once in 1931 and then not again until the 1950s.

**Figure 13 - Sir Douglas Mawson built his research huts at Cape Denison, land of the blizzard. Photo: Mawson's Hut Foundation.**

The patron of the Hut's restoration was Sir Edmond Hillary. I was astonished that he wished to be called just 'Ed' when we chatted during the numerous fund raising events. Billed as the first man to conquer Mt Everest on 29 May 1953, Ed quickly commented that he was part of a team with Tenzing Norgay that climbed the mountain. And the photograph published throughout the world was not him on the summit but Tenzing. "He had never used a camera and the summit of Everest was not the place to teach him how to take a photograph," Ed told me.

While everyone knows about Everest, fewer people know that this great New Zealander, Sir Edmond Percival Hillary, KG, ONZ, KBE, also spent time exploring the Antarctic. He was part of the Commonwealth Trans-Antarctic Expedition in 1958. Using converted Massey Ferguson TE20 tractors, Hillary and his men were responsible for route-finding and laying a line of supply depots up the Skelton Glacier and across the Polar Plateau on towards the South Pole. He reached the South Pole and later in his life would also travel to the North Pole.

Just to complete the picture, in his youth he served as a navigator in the RNZAF during World War II and devoted much of his life to helping the Sherpa people of Nepal, building remote schools and hospitals. What a life!

He died in New Zealand on 11 January 2008, aged 88.

I asked Edmond Hillary if he thought that George Mallory and his partner Andrew Irvine may have reached the summit of Everest in 1924. Mallory made the immortal quote when asked why he wanted to climb Mt. Everest - "because it's there." Ed told me, in typical New Zealander style, "Climbing a mountain involves getting to the top and returning."

In the race to discover the South Pole in 1911, Robert Falcon Scott, *"Scott of the Antarctic,"* became a legendary figure for his iconic polar exploration that resulted in the death of his team on the Ross Ice Shelf during their return after reaching the pole.

Of course, it was Roald Amundsen, a Norwegian explorer who led the first expedition to reach the South Pole. Using skis and dog sleds, he arrived 35 days before Scott's team. Scott was unconvinced that skis and dogs were efficient; he had a preference for man-hauling, pulling the supply sleds themselves - A decision that cost his team their lives in a cold and frozen wilderness.

On our farm in Appin, we own a TE20 tractor. It's close to 60 years old and not only does it still run, but we use it almost every day for farm work. You can still get parts and it starts with the first turn of the key. It seems that Hillary's tractors and Amundsen's dogs still beat man-hauling.

It must have saddened Sir Edmond when, in 2006, Professor Robert Dunbar told the Scientific Committee on Antarctic Research that it is inevitable atmospheric carbon dioxide levels will double and the ice sheets in western Antarctica will be particularly vulnerable. "The ice sheet is grounded way below sea level and that's the part of Antarctica that is most sensitive to global warming over the next few centuries," he said. "When that part melts, and many people think that we are going to see that melt over the next several centuries, sea levels will rise about six to seven meters."[18]

At the same conference, New Zealand Antarctic Research Centre's Professor Peter Barrett highlighted that some 20 million years ago there was very little ice in Antarctica and drill cores show evidence of vegetation. "That is where we are headed," he said.

As recently as 2009, Penn State University's Professor Richard Alley, warned that a melting Antarctica could rise sea levels by 190 feet. Speaking at the American Association for the Advancement of Science, he said, "we don't think that we will lose all, or even most, of Antarctica's ice sheet, but important loses may have already started and could rise sea level as much or more than melting of Greenland's ice over hundreds or thousands of years."[19]

Recent studies also suggest that the ice is melting faster than the projected increased snowfall in Antarctica from the warmer temperatures. This melting ice could even shift the Earth's rotation. Noting that the ice may not melt for centuries, Oregon State University's Professor Peter Clark, claims that the IPCC has estimated that the collapse of the Antarctic ice sheet would result in global sea levels rising by about 5 meters. However, Clark's research shows that several key factors, such as the Earth's rotation and gravity will cause the sea level to rise by 6.3 meters.[20]

The thickest ice measured in Antarctica has been found in Wilkes Land. It reaches a depth of 4776 meters. This is near Australia's Casey Station.

I know this concept is basic, but the basis for all or part of Antarctica to melt is that temperatures rise above the freezing point of water. So where better to start our analysis than at the South Pole itself. The United States Government built the original Amundsen-Scott Research Station on the geographic South Pole during 1956. Dramatically expanded over the years, it is now a massive permanent station. In addition to its remoteness, the South Pole is also elevated 2835 meters above sea level.

Unfortunately, there are too many missing records from this massive facility to perform a comprehensive statistical distribution of temperatures at the base, but it appears from the available data that temperatures at the South Pole have never been above freezing. It was a complete surprise that so much data from an important US base like Amundsen-Scott would be missing from the temperature records.

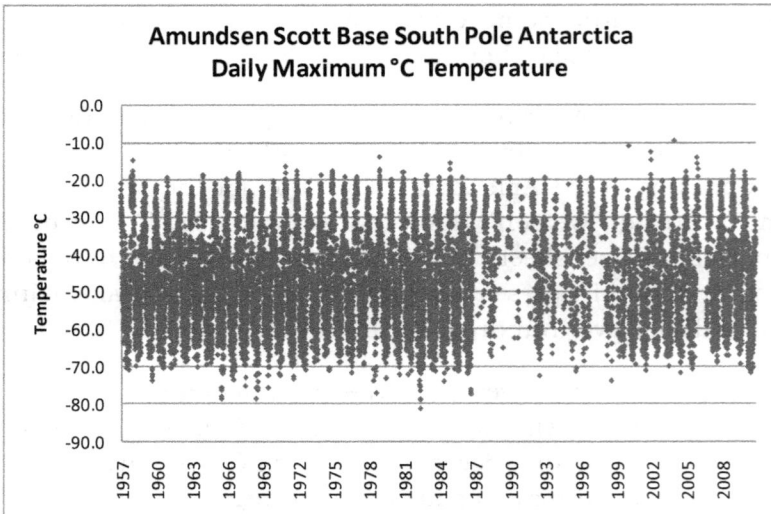

**Figure 14 - Maximum temperatures for the Amundsen Scott Base, located at the geographic South Pole. While 24.1% of the temperature records are missing since 1957, it is clear that the temperature at the pole is well below the freezing point.**

While most bases in Antarctica have weather stations, two locations on the continent have recorded excellent long term readings. These are the Australian stations of Mawson and Davis.

Mawson Station is located on the Antarctic continent's coast in Mac Robertson Land, at the edge of the Antarctic plateau and near the Amery Ice Shelf. It is the longest continuously operating station south of the Antarctic Circle and has temperature records dating back to 1954.[21] While sharing the name of the great Australian polar explorer, Mawson Station is not near Mawson's Huts. The Mawson data shows a very consistent climate. Signs of warming are again not obvious. Refer Figure 16.

Australia's Davis Station is located in Princess Elizabeth Land just across the Amery Ice Shelf. As the penguin swims, just short of 1000 kilometres. It is a remarkable ice free area called the Antarctic Oasis, in the Vestfold Hills. The term oasis is somewhat misleading, as these regions are devoid of snow or ice because they receive almost no moisture. The dryness causes any snow that exists to evaporate from a process of sublimation. You would know this as freezer burn, where a solid transforms itself directly into a gas without the need to melt.

The temperature records for Davis Station are not as good as Mawson. A big chunk of data is missing from 1964-69. Even with the missing data, the daily maximum temperatures from Davis show the temperatures to be very consistent over the years. Refer Figure 17.

The distribution of temperatures for Davis was determined using the full years from 1970-90 and 1991-10, thus avoiding the missing data. This distribution showed some warming at -10°C and 5°C, but no changes to the extremes. Overall however, a very similar temperature distribution, shown in Figure 18, exists over the two samples.

Both these locations show little evidence of global warming.

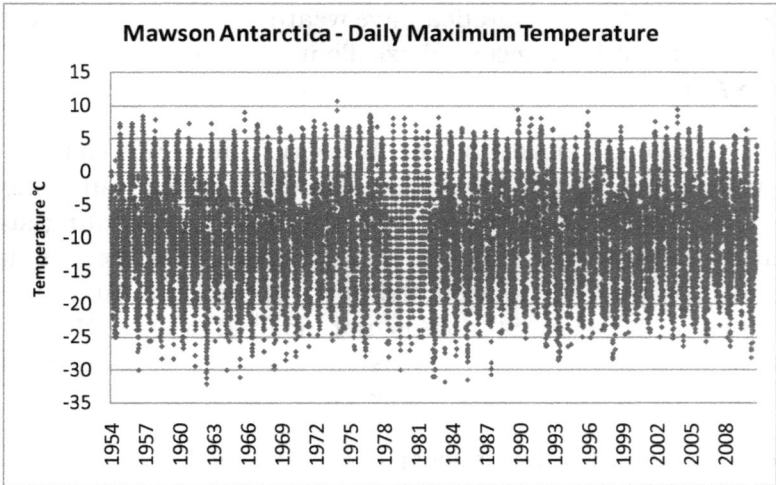

**Figure 15 - The daily maximum temperatures for Mawson Station shows a very consistent climate for over 50 years.**

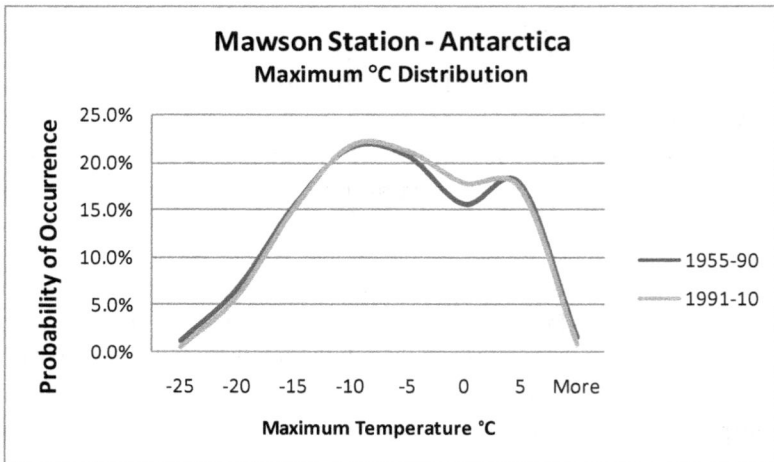

**Figure 16 - The distribution of temperatures from 1955-90 and 1991-10 show almost identical extremes and a slight difference in the camel hump around 0°C degrees.**

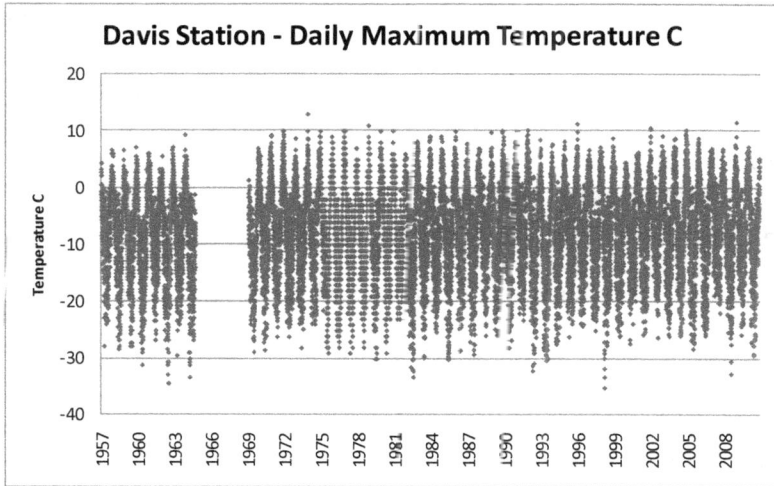

Figure 17 - Daily temperatures for Davis Station also shows a very consistent temperature range. However, the missing data from 1964-69 makes this sample not as solid as the data from Mawson Station.

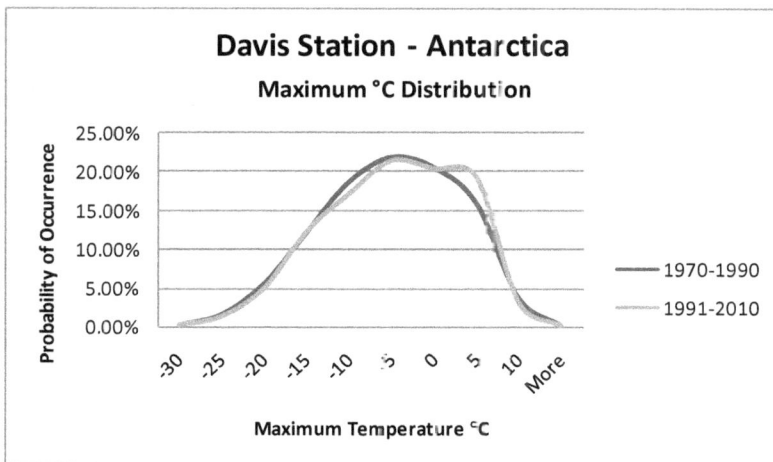

Figure 18 - The statistical distribution for Davis Station shows some changes to the temperatures at -10°C and 5°C however, the extremes are unchanged.

The Australian Bureau of Meteorology should be complimented on the extensive range, audit and sophistication of their Climate Data Online website. The only small issue was the need to download one year at a time to analyse the raw data. This did take a considerable amount of effort since the records extended back for decades. The major limitation with the BOM's data is that it only contains Australian locations including their Antarctic stations.

Finding daily temperature records is challenging. The funny thing is that I am certain that the records were taken.

Even with these challenges, it seems that the authorities are certain that global warming is occurring. As mentioned before, the United Nations Intergovernmental Panel on Climate Change says that the "warming of the climate system is unequivocal."[22] The thesaurus offers us words like *plain, clear, unmistakable, obvious* and *indisputable* for the word unequivocal. Yet the National Climatic Data Centre reminds us that there is no central repository for all the daily meteorological data collected.[23]

The National Climatic Data Centre (NCDC) operates two major sources for daily temperature records. The Global Summary of the Day (GSOD) is the most extensive online database; however it contains quite a few large gaps of missing data and you are lucky if the data starts in the 1970s. The second database from the NCDC is the Global Historical Climatology Network – Daily (GHCN). While there are over 43,000 stations, only 8500 are regularly updated each month with observations. While this might initially sound impressive, compare this to the 44,010 airports in the world.[24] Each of these might have at least some basic meteorological information. There are 15,079 airports in the United States alone. Not to mention all the cities and towns that measure weather. And only 8500 locations are regularly updated.

The GHCN database receives more scrutiny, cross checks and audit than the GSOD. As such, it is considered more accurate.

The British Antarctic Survey READER database contains only monthly and annual averages.[25] From this dataset the assertions that the Antarctic Peninsula had the highest levels of observed climate change in the world were made.

Technically, Orcadas is not really on the Antarctic Peninsula. It is a base in the South Orkney Islands. The Argentines have kept a permanent facility there since 1904 and performed meteorological observations since 1903. The climate records for Orcadas were one of the four stations used to assert the warming claims in the Peninsula; and with its long history, it is the most important.

The GSOD only contains records starting from 1973; very unsatisfactory for such an important location. The GHCN goes back to 1957, but it stops in 2004 for some unknown reason. The monthly READER dataset goes back to 1903. Of importance, most of this dataset was produced by Dr Phil Reid of the Climatic Research Unit at the University of East Anglia. He describes its development from four overlapping datasets as a "nightmare" due to the many discrepancies between them.[26]

Using the two overlapping GSOD and GHCN records, I was able to build a daily history for Orcadas since 1957 up to 2010. Two climate distributions were calculated from 1957-90 and 1991-10. These produced almost identical climates, as shown in Figure 19.

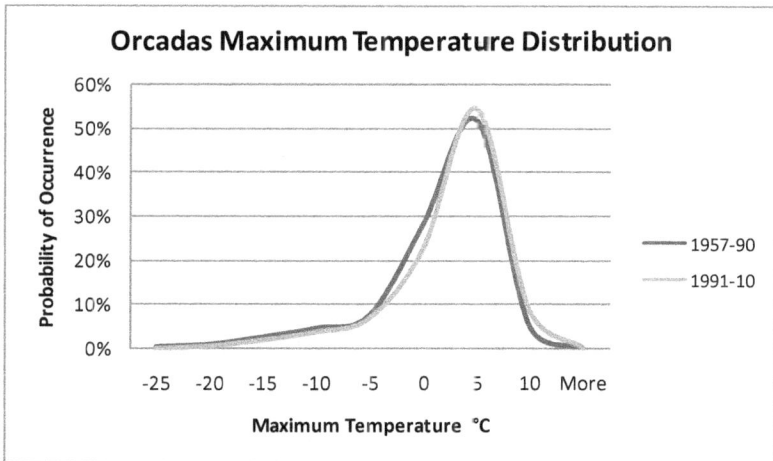

Figure 19 - The two maximum temperature climate distributions for Orcadas are almost identical.

Similar to my findings for Esperanza, the Orcadas data shows little signs of the most extreme warming on the planet. However, a very disturbing finding from the data really shocked me. Maybe I am having the same nightmare as Dr Phil Reid.

Of the 9705 records that overlapped between the GSOD and the GHCN, 3606 records differed by more than ±1°C. That is 37.1% of the sample. And 5% of the sample differed by ±5°C. The largest difference was 19.7°C, when on 16 July 1994, the GSOD was -27.6°C and the GHCN was -7.9°C.

These major differences were not isolated. For example, on 29 May 89, the GSOD records 20.8°C but the GHCN is 1.6°C. Now 20.8°C is a beautiful day. You'll need your long johns at 1.6°C.

What a difference!

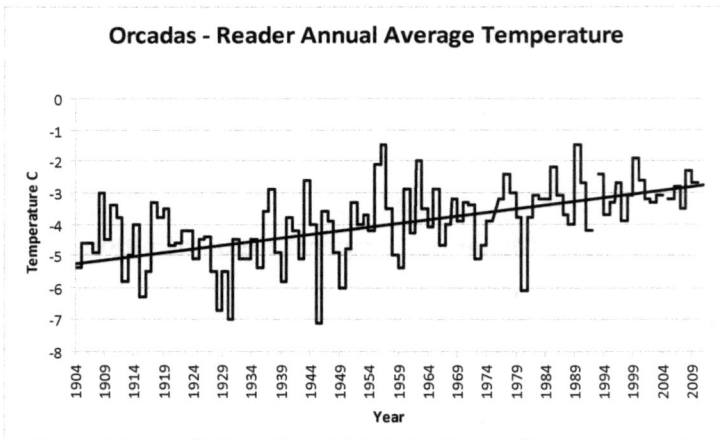

**Figure 20 - Produced by Dr Phil Reid of the Climatic Research Unit at the University of East Anglia, he describes its development from four overlapping datasets as a "nightmare" due to the many discrepancies between them. The linear trend line was used to forecast significant climate change in the region.**

Now, if you were going to make the claim that the warming of the climate system is *unequivocal*, wouldn't you seek to cross check the *plain*, *clear*, *unmistakable*, *obvious* and *indisputable* conclusions drawn from the nightmare of four overlapping datasets with many discrepancies?

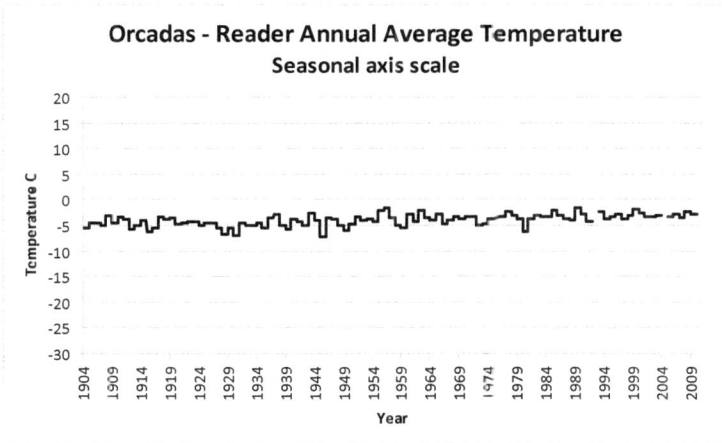

**Orcadas - Reader Annual Average Temperature**
Seasonal axis scale

**Figure 21** –The small axis scale of 0 to -8, used in Figure 20, does not represent the actual climate in Orcadas. When this annual mean temperature is graphed on a scale representative of the range of temperatures on the island, the effect is far less dramatic.

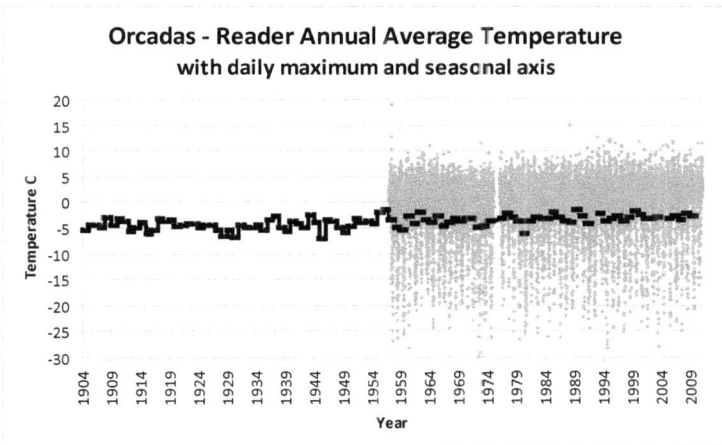

**Orcadas - Reader Annual Average Temperature**
with daily maximum and seasonal axis

**Figure 22** - Finally, the READER annual mean temperature is plotted over the daily maximum temperature. Missing measurements in the daily raw data is obvious.

Like the South Orkney Islands, Macquarie Island is located in the sub-Antarctic. But instead of being isolated between South America and Antarctica at 60°30'S, Macquarie Island lies between New Zealand and Antarctica, at a latitude of 54°30'S. Records for Macquarie Island date back to 1947. They are also very complete, with only 366 days missing in 22,903 days.

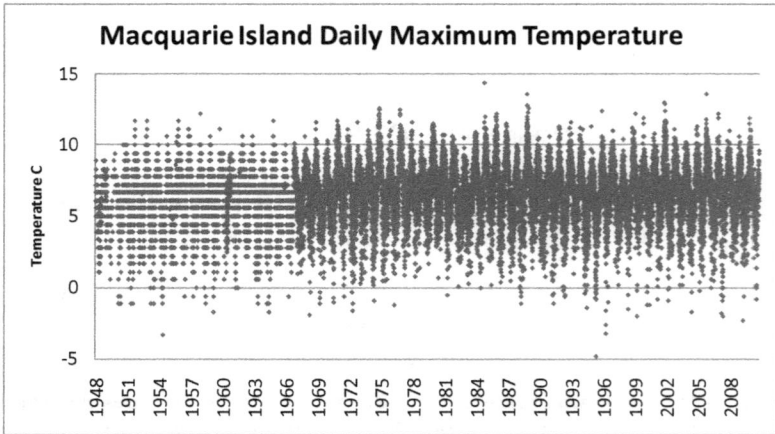

**Figure 23 - The maximum temperatures for Macquarie Island, situated on similar latitude to Orcadas, show no obvious signs of warming. Chapter 6 will address why rounding makes the data looks different from 1948-1969.**

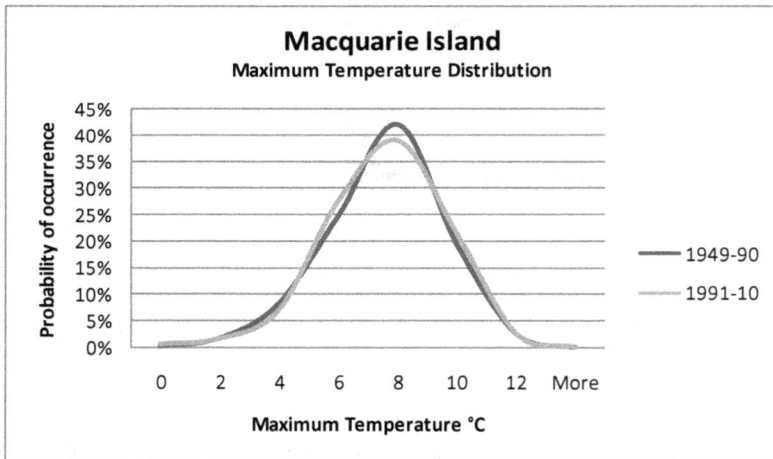

**Figure 24 - Statistical distribution for Macquarie Island shows a very similar distribution, with a slight increase at 7.5°C but no changes to any of the extremes, either high or low.**

Taking this data and placing it into a statistical distribution shows almost an identical climate between 1949-90 and 1991-10, with no changes to the extremes.

Esprenza, Orcadas and Macquarie Island plus the mainland Antarctic bases, certainly do not show *unequivocal* warming trends that could lead to the conclusion of Antarctic ice melt and sea level rise.

The other two locations used in the BAS analysis were Bellinghausen and Faraday. Both of these two datasets have some major limitations for analysis, with missing data. Faraday has 68% of its data missing from 1985-2010. Conclusions from this site are not possible.

Bellinghausen has a sample from 1980 to 1995, but only contains 70% of the days. The days are randomly missing, rather than in large gaps. The second sample is from 1998 to 2010. This sample contains 87% of days. Given these major limitations, the distribution does not support a warming claim and it actually shows a slight cooling trend. (Figure 25)

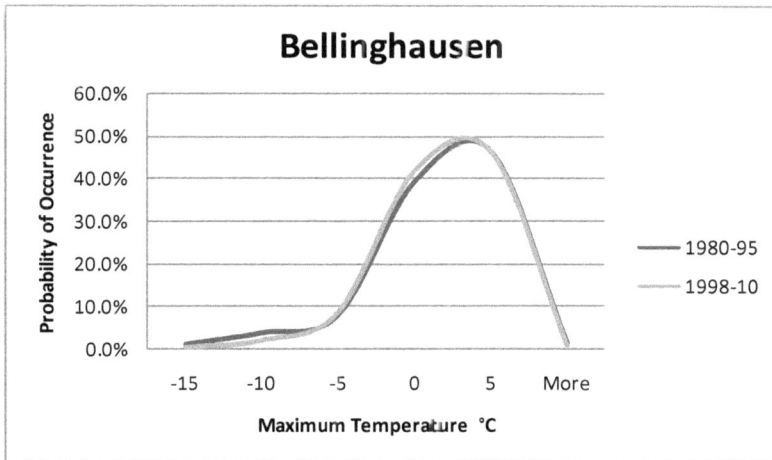

Figure 25 - Bellinghausen was one of the sites used by the BAS to support the warming claims on the peninsula. GSOD temperatures for this location have a considerable amount of missing data, so care needs to be taken in the comparison.

The warming of the Antarctic Peninsula is a keystone in the foundation for the claims made by the UN Intergovernmental Panel on Climate Change. What makes this data even more important is the considerable conclusions that have been drawn from the follow-on effects of warming in Antarctica and in particular the Antarctic Peninsula.

The final thing I would like to show you is what the distribution of temperature looks like if the Peninsula was in fact warming at the claimed rate of 2.5°C.

I took the data for Orcadas and increased every day so we had a difference in the average of the two samples of 2.5°C, the increase in temperature claimed by the BAS and used by the IPCC. I then took this sample and placed the distribution on the graph with the actual temperature measurements.

This example produced a dramatic difference in climates that you would expect from the reports. It is in alignment with the IPCC model (Figure 6) showing less cold days, more warm days and more hot days. However, this is not what occurred. The raw temperature measurements bring the required detail to the analysis.

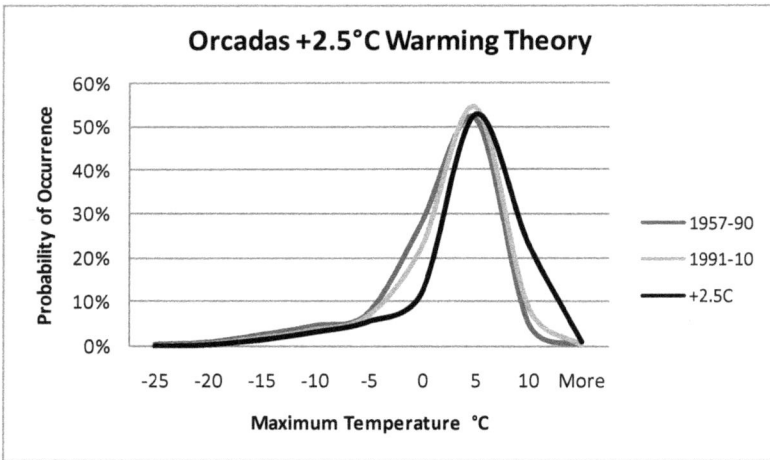

**Figure 26 - An example demonstrating the type of difference that should be observed if the warming claims of +2.5°C based on annual averages translated to daily raw temperature records.**

Remember the reference to the British Antarctic Survey and the climate paper that started it all in 2001 - *Devil in the Detail.*[27]

Unequivocal irony.

# Chapter 5
## LARSEN B ICE SHELF

'But the poles are melting...'

When I was researching this project, I would try to explain that in using the raw temperature data, warming conclusions were invalid. The argument back to me was always the same. 'But the poles are melting.'

Images of melting ice cement the theory of climate change in our minds. Melting ice is everywhere. No single episode had more significance than the coverage of the collapse in 2002 of the Larsen B ice shelf.

At the time, it was claimed that the ice shelf – "larger than the state of Rhode Island" – was the first direct evidence of global warming from human activity. David Vaughan, from the BAS, said "warming on the peninsula has continued and we watched as piece-by-piece Larsen B has retreated. We knew what was left would collapse eventually, but the speed was staggering."[28]

Dr Ted Scambos of the University of Colorado at Boulder, analysed the satellite imagery and found observational proof that ponds of melt water, he assumed were caused by climate change, fractured the shelf by filling cracks in the ice. The weight of the water then drives the cracks through the ice, causing it to shatter.

"Loss of ice shelves surrounding the Antarctic continent could have a major effect on the rate of ice flow off the continent," Scambos notes. "The Ross ice shelf for instance, is the main outlet for the West Antarctic Ice Sheet, which encompasses several large glaciers and contains the equivalent of five meters of sea level in its perched ice."[29]

In 2002, researchers at the Hamilton College Antarctic Program Conference generally agreed that the final straw for the Larsen B Ice Shelf was unusually warm temperatures during the Antarctic summer, December 2001 though to February 2002.

Pedro Skvarca of the Argentine Antarctic Institute said, "Available records show it was the warmest summer since people first brought thermometers to the Peninsula."[30]

Dr Gareth Marshall from the British Antarctic Survey said, "This is the first time that anyone has been able to demonstrate a physical process directly linking the break-up of the Larsen Ice Shelf to human activity. Climate change does not impact our planet evenly. It changes weather patterns in a complex way that takes detailed research and computer modelling techniques to unravel. What we've observed at one of the planet's more remote regions is a regional amplifying mechanism that led to the dramatic climate change we see over the Antarctic Peninsula."

NASA got into the game. Its satellites captured the images of Larsen B's demise and its website states that "Antarctic ice shelves have been shrinking since the early 1970's because of climate warming in the region."[31]

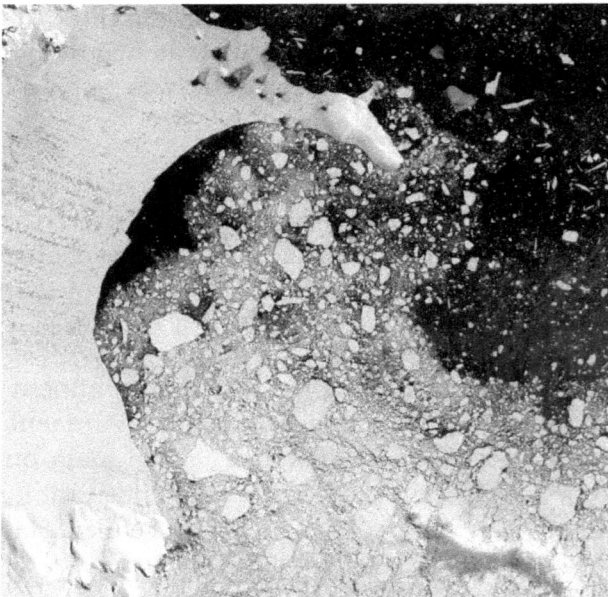

**Figure 27 - Larson Ice Shelf on February 21, 2000. Pools of melt water are visible on the surface of the Larsen Ice Shelf, and drifting icebergs split from the shelf are signs of approaching collapse. (Credit: Image courtesy Landsat 7 Science Team and NASA GSFC)**

The claim is straightforward. Unusual high temperatures, from global warming in the region caused the ice to melt into surface ponds. This water worked its way into the cracks in the ice and this fractured the shelf.

What is certain is that the ice shelf collapsed. Unusual high temperatures, not so certain.

Marambio base is on Seymour Island, just north of Snow Hill Island at 64°14′S 56°37′W. It is on the east coast of Antarctica, around 250 kilometres north of the Larsen Ice Shelf.

As shown in Figure 28, summers in Marambio are quite volatile. There is no such thing as a typical summer's day along the wild coast of the Weddell Sea.

Temperature measurements for each Southern Hemisphere summer, from the start of September to the end of March, were taken from the GSOD from 1973-89. This was compared to the summer of 2001-02, shown in black in Figure 28.

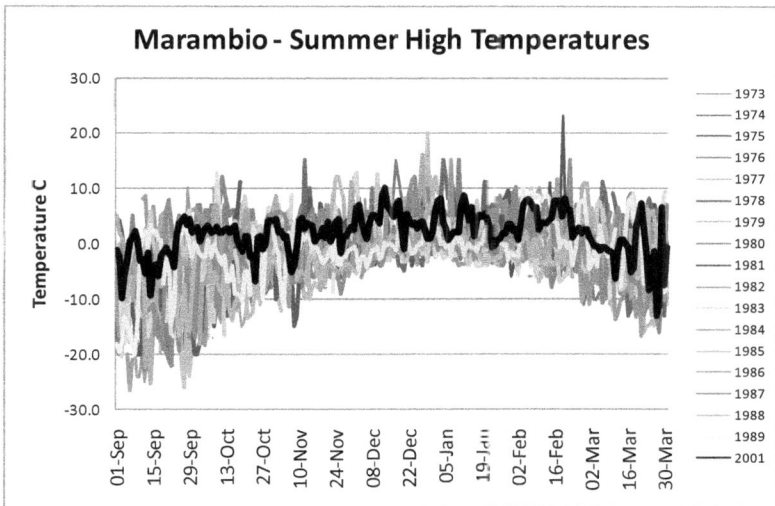

**Figure 28 - Temperatures at Marambio, north of the Larsen B ice shelf, show considerable volatility but are not unusual in terms of warmth occurring prior to the ice shelf collapse starting in January 2002 and lasting 35 days. The summer prior to the collapse is shown in black. The highest temperature of 23°C occurred on 19 February 1974.**

There was nothing unusual about the summer temperatures in the region near the Larsen B Ice Shelf. It is clear that unusual high temperatures caused by global warming did not cause the Larsen B ice shelf to collapse.

But don't just take my research to prove this. In 2008, in a paper published in the Journal of Glaciology by Professor Neil Glasser of Aberystwyth University and Dr Ted Scambos, who's initial research contributed to the warming claims, now at the University of Colorado's National Snow and Ice Data Centre, rebutted the initial theory.

"The ice shelf collapse is not as simple as we first thought," said Professor Glasser, lead author of the paper. "Because large amounts of meltwater appeared on the ice shelf just before it collapsed, we had always assumed that air temperature increases were to blame. But our new study shows that ice-shelf break up is not controlled simply by climate. A number of other atmospheric, oceanic and glaciological factors are involved."[32]

Larsen B did not collapse from global warming.

# Chapter 6
## THE DIGITAL ERA

The first thing I did with my new found income after graduating from Penn State was to take up flying. I was living in Poughkeepsie, New York. Our regional airport was called Duchess Country, where I had my first few lessons. After I pre-flighted the Cessna 152, a two seat trainer built about the same year that I was born, I climbed in and started the engine. Tuning the radio onto the ground frequency, I made my first radio call and promptly became very confused when I heard the response.

> "Cessna six four two niner bravo, taxi active and hold, barometer two niner niner six, wind two six at twelve."

I thought to myself that I was never going to understand these radio calls.

After I completed about 10 hours of flying, I moved to a tiny airport called Sky Acres. It was a small, uncontrolled regional airstrip and I was told that I could get in more flight hours because I didn't have to spend as much time taxiing and holding.

Sky Acres had its own set of particular features. The runway ran uphill and if you landed on the numbers painted on the runway to tell you its direction, where you were supposed to land I might add, you hit a sink hole that bounced the aircraft back into flight. Like many regional strips, you had to hit the microphone key five times to turn the on the runway lights at night. But since you had to buzz the runway first to make sure there was no wildlife – from deer to cows – that would make a real mess of your aircraft, the lights had a habit of turning off just as you were on final approach.

This might be why the locals called it Scare Acres.

I had a great flight that I loved to do. The people that flew it with me still talk about it 25 years later. Taking off at dusk, I would fly over the Roosevelt and Vanderbilt mansions that dotted the Hudson River. A sweeping turn over *"The Gunks"*, a beautiful rock climbing Mecca where I spent my time when I wasn't flying, put us on a flight path down the Hudson River. A quick wing over as we passed West Point Military College. With their collection of historic guns and ramparts, you could appreciate the skill of a bombing mission in World War II. It's getting dark now, and the final brilliant colours of the sunset are fading. The lights of New York City can be seen in the distance. Just after you fly over the Tappan Zee Bridge, you lower the altitude to 500 feet and you follow the Hudson River through the lights of Manhattan. The buildings are all taller than your flight altitude.

Then you see the USS Intrepid. A once proud aircraft carrier, commissioned in 1943, she saw action from the battle of the Leyte Gulf in World War II to Vietnam. She was also the aircraft carrier that recovered the Mercury and Gemini space capsules. She is now an aircraft museum in New York City and looks so tiny from the air. You thank God that you get to land at Sky Acres rather than on her deck in rough seas with a Japanese Zero on your tail.

But then you see the other woman that grabs your attention – the Statue of Liberty. You get to make a circle just next to her, the view was magnificent.

These flights were in 1985. Unfortunately, following 911 this airspace is now closed. This flight will only be a memory for a few of us now aging pilots.

Another memory of flying is seeing a person peering into a small structure to record the weather. Weather reports are the foundation of light aircraft aviation. One day I had a peek inside. It wasn't exactly high technology.

It had just a few instruments.

> A thermometer, barometer and hygrometer; for measuring temperature, barometric pressure and humidity. An anemometer; to determine the wind speed. A wind vane; for measuring its direction. And a rain gauge, to measure precipitation.

Twenty years later, my best friend Brian from Penn State came out to Australia for an adventure. He was now a pilot for the airlines and the two of us were going to rent a light aircraft and fly to the Great Barrier Reef. I hadn't flown in almost a decade and couldn't believe how well I landed the plane during my check ride. I guess it must be like riding a bike.

This time we rented a Cessna 172, a four seat version of the 152. The aircraft hadn't changed in the 20 years. It was of the same vintage as the aircraft I trained in and it was still almost as old as I was. The flight instruments were still all analogue, with hands or dials that moved. But what did change is that we now carried a handheld Global Positioning System (GPS) to track our flight. We had entered the digital age. This thing was so accurate; we never missed a check point during the thousands of kilometres we flew over the Australian outback.

The other thing that changed in the past decade is the humble thermometer. Instead of being filled with mercury and being read by the fluid level, it is now also digital.

In his book from 1850, *A Manual of the Thermometer*, John Henry Belville, of the Royal Observatory, Greenwich, provided a detailed history into the development of this instrument.[33]

> In the beginning of the 17th Century, Drebbel of Alkmaer and Sanctorio, a professor of Padua both claim the invention. These inventions used a coloured liquid in a blown glass tube which expanded and contracted as the temperature changed. In 1701, Sir Isaac Newton filled the tube with linseed oil, but the viscidity caused it to adhere to the sides of the tube. In 1724, Fahrenheit filled the tube with mercury.

So we would have measured temperature in Newtons if Sir Isaac would have used mercury instead of linseed oil.

Again from Belville's book, the other problem that proved difficult in the invention of the thermometer was determining:

> The fixed point from where the scale would commence. Celsius adopted the freezing of water as zero and the boiling of water as 100. Fahrenheit made his zero from the mixture of snow and common salt, he divided the interval between that and the freezing point of water into 32 parts and the interval between the freezing and boiling of water into 180 parts.

John Belville also offered some interesting instructions for using the thermometer. "The observations should be read off quickly, more particularly in cold weather; the observer should be careful neither to touch or breathe, or indeed approach the body too closely to the instruments. The eye should be placed on level with the mercury, and accuracy should be the animus of observation."

The word *animus* came into use about 1815. It means purpose or passion.

Would John Belville ever of dreamed that in 2011, his book would be digitised by Google and be accessible by almost everyone on the planet? Welcome to the digital era. And while John was very passionate about the thermometer, he even measured altitude with his thermometer, using the changing temperature of boiling water on various mountain peaks; I don't think he ever would have dreamed that his measurements were critical to the future of the planet.

Yet John clearly understood the importance of temperature and climate measurements. His preface talks about just how variable the climate of England can be:

> Having entered in the year 1811 on duties at the Royal Observatory, which required attendance not only by day but also by night, my attention was necessarily very early directed to the great changes to which the climate of England is susceptible: the sudden formation and disappearance of the clouds, the great variations of temperature, the unsteadiness of the mercurial column, and its connexion with the different winds which blow, became the subjects of interest, and were tabulated day by day until they have now accumulated into a mass of observation from which much valuable information is deduced.

I'm not certain that since the invention of the thermometer, the tens of thousands of people recording the temperature each day in weather stations around the globe were *animus of the observation.*

However, the world's climate scientists don't use these raw observations to determine if the climate is changing. They convert the temperatures to determine what they call *'anomalies'* against a long term average from 1961 to 1990. This is their way to compensate for the vast amount of missing data. Dr Peter Stott, Climate Monitoring Expert from the UK Met Office explains this on their website:[34]

> Absolute temperatures are not used directly to calculate the global-average temperature. They are first converted into 'anomalies', which are

the difference in temperature from the 'normal' level. The normal level is calculated for each observation location by taking the long-term average for that area over a base period. For HadCRUT3, this is 1961–1990.

For example, if the 1961–1990 average September temperature for Edinburgh in Scotland is 12 °C and the recorded average temperature for that month in 2009 is 13 °C, the difference of 1 °C is the anomaly and this would be used in the calculation of the global average.

One of the main reasons for using anomalies is that they remain fairly constant over large areas. So, for example, an anomaly in Edinburgh is likely to be the same as the anomaly further north in Fort William and at the top of Ben Nevis, the UK's highest mountain. This is even though there may be large differences in absolute temperature at each of these locations.

The anomaly method also helps to avoid biases. For example, if actual temperatures were used and information from an Arctic observation station was missing for that month, it would mean the global temperature record would seem warmer. Using anomalies means missing data such as this will not bias the temperature record.

I can't believe how far we are from my sister's recommendation to use the raw data.

*HadCRUT3* is described as a gridded dataset of global historical surface temperature anomalies for each month since January 1850. The dataset is a collaborative product of the Met Office and the Climatic Research Unit at the University of East Anglia.

The Met Office and its association with the Climate Research Unit were brought into the global spotlight when thousands of emails were made public following what Government officials deemed as illegal hacking. The media dubbed the leak *"Climategate"* and allegations were made of misconduct within the climate science community. A subsequent panel investigated the unit. [35] Its members were all academics.

**International Panel to examine research of Climate Research Unit**

Chair: Prof Ron Oxburgh FRS (Lord Oxburgh of Liverpool)
Prof Huw Davies, ETH Zürich
Prof Kerry Emanuel, Massachusetts Institute of Technology
Prof Lisa Graumlich, University of Arizona.
Prof David Hand FBA, Imperial College, London.
Prof Herbert Huppert FRS, University of Cambridge
Prof Michael Kelly FRS, University of Cambridge

Their terms of reference were very specific. They were to assess the integrity of the research published by the Climatic Research Unit in the light of various external assertions. The team was not concerned with the question of whether the conclusions of the published research were correct.

The report's conclusions are offered here in full:

1. We saw no evidence of any deliberate scientific malpractice in any of the work of the Climatic Research Unit and had it been there we believe that it is likely that we would have detected it. Rather we found a small group of dedicated if slightly disorganised researchers who were ill-prepared for being the focus of public attention. As with many small research groups their internal procedures were rather informal.

2. We cannot help remarking that it is very surprising that research in an area that depends so heavily on statistical methods has not been carried out in close collaboration with professional statisticians. Indeed there would be mutual benefit if there were closer collaboration and interaction between CRU and a much wider scientific group outside the relatively small international circle of temperature specialists.

3. It was not the immediate concern of the Panel, but we observed that there were important and unresolved questions that related to the availability of environmental data sets. It was pointed out that since UK government adopted a policy that resulted in charging for access to data sets collected by government agencies, other countries have followed suit impeding the flow of processed and raw data to and between researchers. This is unfortunate and seems inconsistent with policies of open access to data promoted elsewhere in government.

4. A host of important unresolved questions also arises from the application of Freedom of Information legislation in an academic context. We agree with the CRU view that the authority for releasing unpublished raw data to third parties should stay with those who collected it.

Let's take a moment to understand these conclusions. They performed their work without close collaboration with professional statisticians. There are important and unresolved questions regarding the environmental datasets. To me, these are very disturbing findings.

To understand what the important and unresolved questions regarding temperature are, you only have to look to page 3 of their report:

> These records are irregularly distributed in space and time. Modern records come largely from land-based meteorological stations but their geographical distribution is uneven and strongly biased in favour of the northern hemisphere where most of the Earth's land masses are located. Oceans cover two thirds of the Earth's surface and away from the main shipping routes coverage is thin. For earlier centuries the record is much sparser. Deriving estimates of past temperatures on a global, hemispheric and regional scale from incomplete data sets is one of the problems faced by the Unit and in consequence an important current interest is the discovery of useable old temperature records from a variety of sources.

At this point I formed my own conclusion. We have based the entire model for climate change on incomplete datasets with '*anomalies*' derived from averages and missing data, ignoring the oceans, without the input of professional statisticians.

Our society has a perception that there is massive Government and scientific resource assigned to analyse data and determine the policy for community good. The Climate Research Unit at the University of East Anglia consists of only 3 full time and 1 part staff members supported by about 12 PhD students and assistants.

They were described by the investigating panel as disorganised and unprepared for the international media spotlight. They were praised for performing the "time consuming and meticulous" work sorting through old temperature records when it was "unfashionable" to do so. The CRU were also concerned that their data and conclusions were used incorrectly by the IPCC and the WMO, as described in this extract from the findings:

> Recent public discussion of climate change and summaries and popularizations of the work of CRU and others often contain oversimplifications that omit serious discussion of uncertainties emphasized by the original authors. For example, CRU publications repeatedly emphasize the discrepancy between instrumental and tree-based proxy reconstructions of temperature during the late 20th century, but presentations of this work by the IPCC and others have sometimes neglected to highlight this issue.

The panel also noted that that the published work from the CRU contains many cautions about the use of their data. The exact quote from the report is as follows:

> The published work also contains many cautions about the limitations of the data and their interpretation.

On 3 December, to coincide with the 16th session of the *Conference of Parties of the United Nations Framework Convention on Climate Change,* held in Cancun, Mexico, the World Meteorological Organization released the results that "over the ten years from 2001 to 2010, global temperatures have averaged 0.43°C above the 1961-1990 average, the highest value yet recorded for a 10-year period." These warming claims were extensively reported by the world's media as direct evidence of climate change.

Reference was made by the WMO that the data to make this conclusion was sourced from the Climate Research Unit. Yet not one of the caveats, none of the cautions about the limitations of the data and their interpretation, were referenced in the WMO's press release. The dust was not even settled on *Climategate* and the cautions were ignored.

This investigative report is an excellent example of the abuse of information, but the guilty parties are the IPCC and the WMO rather than the disorganised researchers at the Climate Research Unit.

Weather stations around the world are being closed from budget pressure and the Climate Research Unit, the keeper of the world's weather data, has only 3 full time staff. Historical temperature records have been described by the creator as a nightmare. The findings are popularised and often contain oversimplifications. The governing bodies, including the IPCC itself, ignore these discrepancies.

Surely the World Meteorological Organisation saw the report by the International Panel to investigate the CRU. Obviously they were aware of the caveats, cautions and limitations of the data. To be told in the report that the data has been popularised and oversimplified should have sent alarm bells ringing within the WMO that their science and audit capability needed to be seriously reviewed.

But no, their response was to exploit the situation with yet another oversimplification, "the highest value yet recorded for a 10-year period."

Do you think I'm being too critical? Remember what they were measuring, the difference between the years 1960-1990 and the 10 years from 2001-10. There has only been one ten year period to compare it to, 1991-2000. I've worked in media communications; I know how to spot a headline chaser. The World Meteorological Organisation is supposed to be a science organisation.

So now let us go back to the lonely thermometer of John Belville. Between 1724 and 1990, when these temperatures were read from a bulb thermometer, it was often common practice to round the temperature down when measurements were taken. Also, temperatures were taken at intervals, most often on the hour, so actual maximums and minimums may never have been recorded. Some stations had special thermometers that caught the maximum and minimum, but many weather stations just had a basic thermometer.

The GHCN calls this bias adjustment, "resulting from historical changes in instrumentation and observing practices." While they acknowledge the problem, no attempt has been made to adjust temperatures for these factors.

After 1990 we began the digital age. The bulb thermometer was replaced by a digital thermometer. This was more accurate and now records temperature in tenths of degrees. Weather stations were automated. Maximum and minimums are kept in computer memory and are never missed.

So a hot day in 1963 might have been 32°C degrees, but in 2007 it might have been 32.6°C degrees. If the maximum didn't occur during the hour sampling period, it may have even been warmer.

Often you can visualise the bias from the simple when you are examining the raw data. An example of this is for Tasiilaq, Greenland. Taken from the GHCN database, the table below shows temperatures in 1973 were rounded but in 1996, the digital measured temperatures are clear.

| Tasiilaq, Greenland - Maximum °C | | | |
|---|---|---|---|
| 02-Dec-73 | 5 | 02-Dec-96 | -0.8 |
| 03-Dec-73 | 0 | 03-Dec-96 | -4.2 |
| 06-Dec-73 | -5 | 06-Dec-96 | -2.6 |
| 08-Dec-73 | 0 | 08-Dec-96 | -7.8 |
| 10-Dec-73 | -10 | 10-Dec-96 | -9.2 |

The Australian Bureau of Meteorology calls this issue a homogenous climate record:[36]

> A change in the type of thermometer shelter used at many Australian observation sites in the early 20th century resulted in a sudden drop in recorded temperatures which is entirely spurious. It is for this reason that these early data are currently not used for monitoring climate change. Other common changes at Australian sites over time include location moves, construction of buildings or growth of vegetation around the observation site and, more recently, the introduction of Automatic Weather Stations.

The two temperature scales developed by Celsius and Fahrenheit may also have caused differences in the historical global climate record. Personally, having lived under both systems, each does its job adequately.

However, simply because there are more units on the Fahrenheit scale, it would have been more accurate than the measurements taken in Celsius when it came to reading bulb thermometers.

This photograph is of the thermometer on our farm. It is located in a shaded area just outside our back door. It was taken on the hottest day we experienced since moving onto the land.

It is 110°F or 43°C.

**Figure 29 - Our farm thermometer showing our warmest day of 110F or 43°C. Or should it have been 43.3°C.?**

Or should it have been 43.3°C? Perhaps it was only 109°C, which is 42.7°C. It clearly didn't get to 44°C. But it might have 10 minutes later while we were not watching.

Was I animus of observation?

A digital thermometer in an automated weather station would have captured the data perfectly.

Too bad we didn't have automated stations until recently. Those poor souls would not have had to climb out of their sleeping bags on those cold Antarctic days each hour.

Wouldn't it be ironic if the increase announced by the WMO of 0.43°C was simply the result of rounding error brought on from the digital era?

# Chapter 7
# THE END IS NIGH

About 45,000 people travelled to Copenhagen in 2009 with high hopes of a global agreement to combat climate change. However, at the end of the summit, with the convention blanked by newly fallen snow and freezing temperatures, the world's leaders failed to reach any agreement.

Gervais Marcel, from Valetta in Malta, commented on the BBC's blog of the event.[37]

> "The Copenhagen summit on climate change failed for only a simple reason: the lack of political will to understand, recognise and accept the scientific evidence and the worldwide daily havoc that is testimony of the reality of climate change."

For anyone that even pays the slightest attention to the news, reports on the impact of climate change are impossible to miss. Heatwaves and rising ocean levels dominate the media landscape. A few examples:

NewScientist magazine reported that the European heatwave of 2003 caused 35,000 deaths and we should expect more such events from climate change. "The Earth Policy Institute says it is confident that the August heatwave has broken all records for heat-related deaths and says the world must cut the carbon dioxide emissions that contribute to global warming."

Saufatu Sopoanga, former Prime Minister of Tuvalu, told the 58th session of the United Nations General Assembly in New York, on 24 September 2003 that "we live in constant fear of the adverse impacts of climate change."

In the Pacific Islands Ministerial Conference on Environment and Development in Asia and the Pacific 2000, the topic of sea level rises from climate change was discussed. Data presented at the conference and compiled from the 11 tide gauges operated by Australia's National Tidal Facility at Flinders University, concluded "that changes in sea level are related to a multitude of variables and no realistic trend can be detected from the data for many years to come."[38]

That might not have been the most political remark. In 2004, the National Tidal Facility was closed and its function transferred to the Bureau of Meteorology.

The same conference was told that after 68 months of operation, the average sea level rise in Samoa turned out to be a sea level fall of -19.7mm per annum.

Global warming has been identified as destroying coral reefs from the rising sea levels and bleaching, which occurs in warmer water. Yet the *Samoa News* reported in 2010 that researchers have found that some South Pacific coral atolls have even grown in size over the past 60 years.[39]

Reports are widespread: Rising and acidic sea levels, melting glacier ice, glacier retreat, warmer seawater, Arctic, Greenland and Antarctic ice thinning or melting, increased rainfall causing major flooding in some parts of the world and record drought in others. Polar bears will become endangered while scores of other species will have no home.

Hurricanes will grow in frequency and strength. All this increased moisture will result in a dramatic increase in mosquitoes and disease, not to mention an increase in pollen and allergy attacks.

But if you think this is climate change Armageddon, just look what Reuter's reported in 2009. Earth quakes, volcanic eruptions, giant landslides and tsunamis may become more frequent as global warming changes the earth's crust.[40] It seems that the removal of the weight from the melting ice will allow the Earth's crust to bounce back to its original shape, causing these catastrophes.

It's no wonder that Gervais Marcel believes we are seeing the effects of climate change every day. We may be. Or are we just seeing the changing and dynamic weather the Earth as experienced since its formation.

I've got a great book called the *Chronicle of Australia*. It presents a history since European colonialism in newspaper articles. It seems we have been blessed with a few headline events over the past 200 years - War, Prime Ministerial resignations, miner's strikes, droughts, bush fires, floods. It happened in 1810, 1910 and it is still happening in 2010. For example:

> In 1792, Parramatta was severely damaged by storm and in 1795, Hawkesbury river floods were both severe and hit at alarming speed. The floods again devastated the Hawkesbury farms in 1799, again in 1806 and in 1809. In 1820 a caterpillar plague hits the farms around Sydney. By 1852 a raging flood takes 77 lives at Gundagai. In 1893 Brisbane became a drowned city after the worst flooding since settlement and in 1894, two hurricanes swept through Broone. Three hundred people died when one of the worst cyclones in recorded history flattened Cooktown in 1899. By 1902 rains were relieving the drought in Tasmania and in 1903 the drought broke in Broken Hill. In 1906, floods in New South Wales, Victoria and Tasmania were the worst for 20 years while bushfires claimed 13 lives in Gippsland.

All this before the start of the First World War.

The point I am trying to make is where do you draw the line between a normal weather event and a climate change event?

The UK climate change office says London will experience wetter winters and more frequent heavy downpours. Yet since 1948, the rainfall for Heathrow Airport shows no change to London's rainfall patterns.

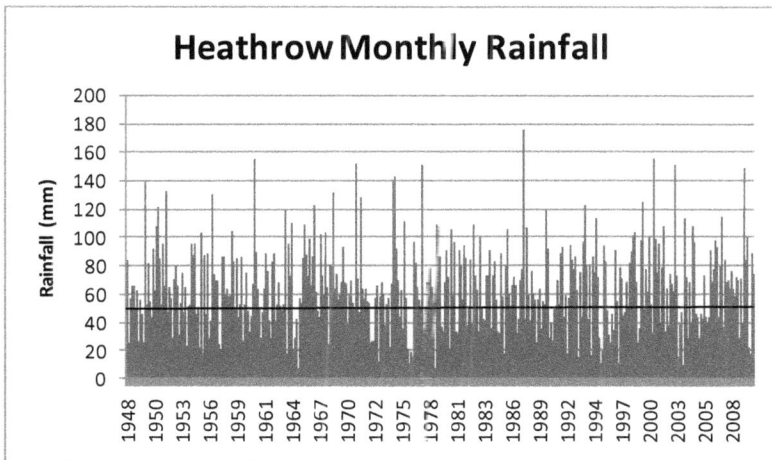

**Figure 30 - Monthly rainfall in London as measured at Heathrow Airport. Consistent rainfall has occurred in London since 1948. Source: UK Met Office.**

When Hurricane Katrina struck the city of New Orleans in 2005, it was reported that the 2005 Atlantic hurricane season, influenced by climate change, was the most active in recorded history. Following the devastation in the city, scores of media reported that Katrina was the creation of global warming.

Yet hurricanes were far from unique to New Orleans.

Louisiana has been hit by 49 of the 273 hurricanes that made landfall between 1851 and 2004.[41] That is 18%. On average, one major storm crosses within 100 miles of New Orleans every decade.

Since New Orleans sits in a depression 1.8 meters below sea level, the destruction of the city was not from the Hurricane, but from the flooding which occurred when many of the levees protecting the city were breached. Even before the storm, the US Federal Emergency Management Agency listed a hurricane strike in New Orleans as one of the most serious threats to the nation.

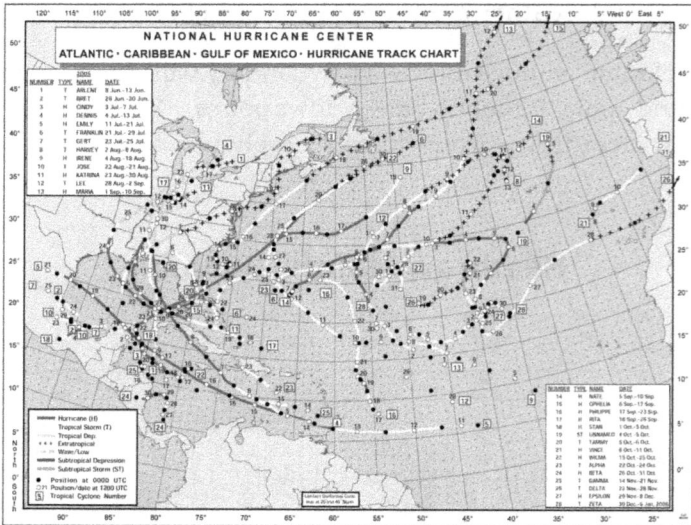

**Figure 31 - The 2005 Hurricane season was billed as the worst in recorded history and offered as proof to climate change. Source: www.nhc.noaa.gov**

While Katrina was a major storm with incredibly low pressure of 920 millibars or 27.17 inches of mercury, as it hit landfall, it was only a Category 3 storm, defined in the US as winds between 111 –

130 mph with a surge of 9 to 12 feet, expected to cause extensive damage.

In 1935, even before hurricanes had sexy names, Hurricane FL was the most intense storm to hit the United States. It had pressure of 892 millibars and was a Category 5 storm - *winds greater than 155 mph, surge over 18 feet and catastrophic damage.*

Measuring hurricanes since 1851, the US has averaged 17.9 hurricanes per decade. In reviewing these storms, the National Hurricane Centre concluded *"the United States hasn't seen a significant resurgence of exceptionally strong hurricane landfalls."*[42]

It also concluded that a Category 4 or stronger hurricane "strikes the United States about once every 7 years" and the US has not seen more hurricanes or increased storm intensity from global warming.

In Australia, where hurricanes are called tropical cyclones, the same type of information is available from the Australian Bureau of Meteorology.[43] On their website, you can enter a year range and track cyclones in the Australian region.

The data for 15 years from 1961-76 and 1991-06 shows no increase in cyclone activity. Refer Figure 34. Have a play on the website yourself; there is no obvious change since records existed from 1907.

After acknowledging a high level of interest on the impact of tropical cyclones and climate change, Dr Geoff Love, the Australian Director of Meteorology, submitted a paper on the subject to the World Meteorological Organisation.[44] They also issued a media release on 20 February 2006.[45] It said:

"No single high impact tropical cyclone event of 2004 and 2005 can be directly attributed to global warming, though there may be an impact on the group as a whole."

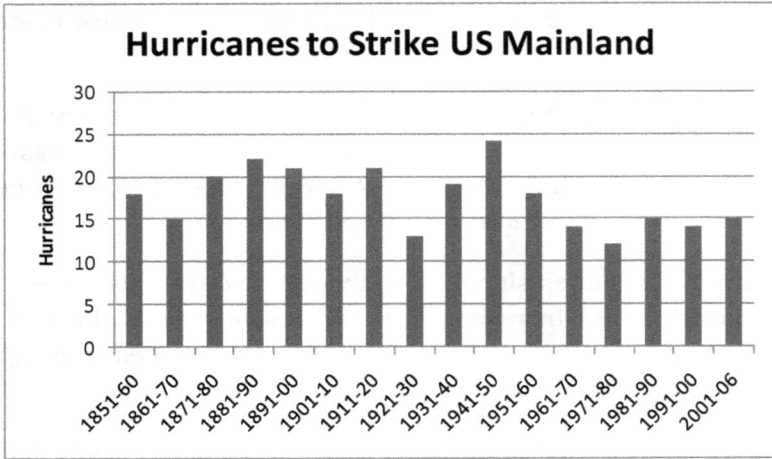

**Figure 32 - Total hurricanes each decade to cross into the US mainland.**

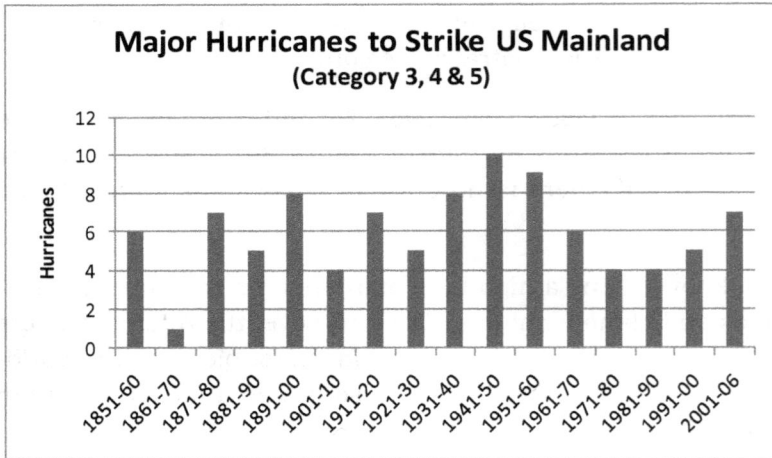

**Figure 33 - Major hurricanes, category 3, 4 and 5, to cross into the US mainland for each decade.**

Tropical Cyclones
1961 to 1976

Australian Government
Bureau of Meteorology

Tropical Cyclones
1991 to 2006

Australian Government
Bureau of Meteorology

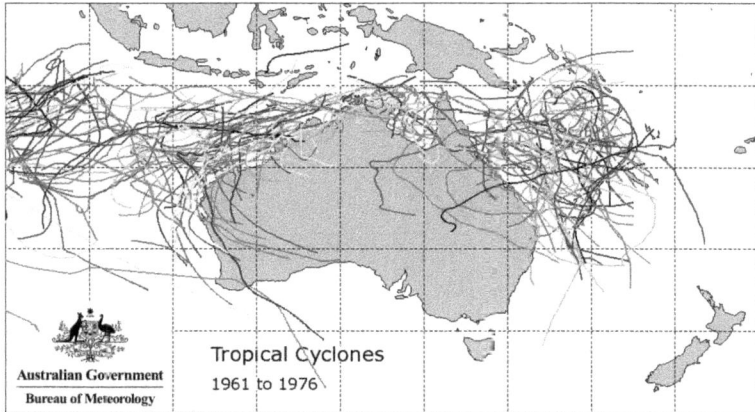

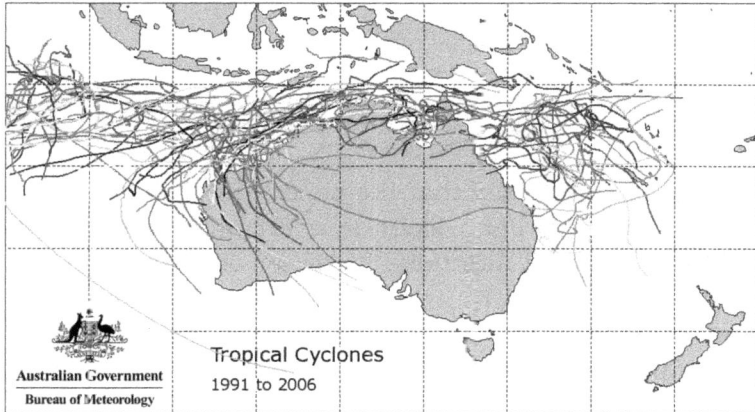

**Figure 34 - The top image shows the tropical cyclones for 15 years from 1961 to 1976 contrasted to the bottom image of cyclones from 1991-2006. Based on this data, the number of cyclones appears to be static or slightly decreasing, not increasing in volume as forecast by climate change.**

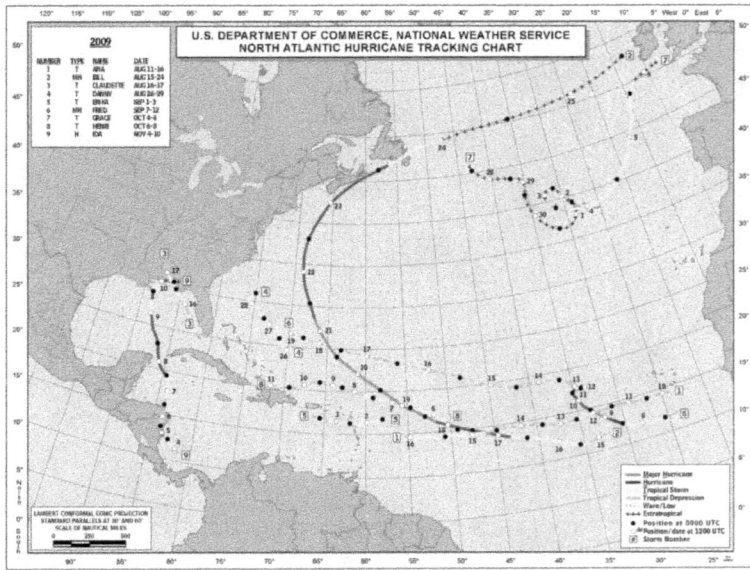

**Figure 35 - The 2009 Hurricane Season saw only 3 hurricanes make US landfall. Source: www.nhc.noaa.gov**

The release goes on to say that the paper was prepared by a group of experts comprising John McBride and Dr Jeff Kepert of the Bureau of Meteorology in Australia, Professor Johnny Chan of China, Julian Heming of the UK, and Dr Greg Holland, Professor Kerry Emanuel, Thomas Knutson, Dr Hugh Willoughby and Dr Chris Landsea of the US.

Two significant observations that the group made were:

"The paper reaffirms the finding of a 1998 study saying that any change in the frequency of tropical cyclones (hurricanes/typhoons) due to climate change cannot be determined due to a lack of knowledge and limitations of the available observing technologies. The little evidence that does exist indicates little or no change in global frequency."

"No single disaster caused by a tropical cyclone (hurricane/typhoon) in 2004 or 2005 - including Hurricane Katrina in the US - can be directly attributed to global warming. Rather, climate change may have an impact on the group as a whole."

Of interest, the group was unable to reach a conclusion on the topic of increased intensity of tropical cyclones. Citing inconsistent climate records and changes in observations and methods over time, the panel says it cannot come to a definitive conclusion in this "hotly debated area."

While this release needs to be applauded for the need to keep balance in science, its coverage in the media was minimal and reports continue to describe 2005 as the worst hurricane season in history.

Katrina was the most costly hurricane to hit the US. However, this says more about the poor quality of levee construction and city complacency rather than the impact of climate change.

The media is also silent about hurricane seasons since 2005. For example, in 2009, the US saw only 3 hurricanes for the season.

Weather events happen. Human nature desires the current event to be bigger and worse than in the past. The use of hurricanes, cyclones and typhoons are again part of the popularisation issue raised by investigators into the Climate Research Unit. In science, it is vital to review all the facts rather than promote the facts that fit the theory and ignore the facts that don't.

Maybe the end is not so nigh?

# Chapter 8
## CARBON MEASUREMENTS

If all roads lead to Rome, then all $CO_2$ measurements lead to Mauna Loa, Hawaii. This station is part of the U.S. Department of Commerce, National Oceanic & Atmospheric Administration, Earth System Research Laboratory, Global Monitoring Division. It is a bit of a mouth full.

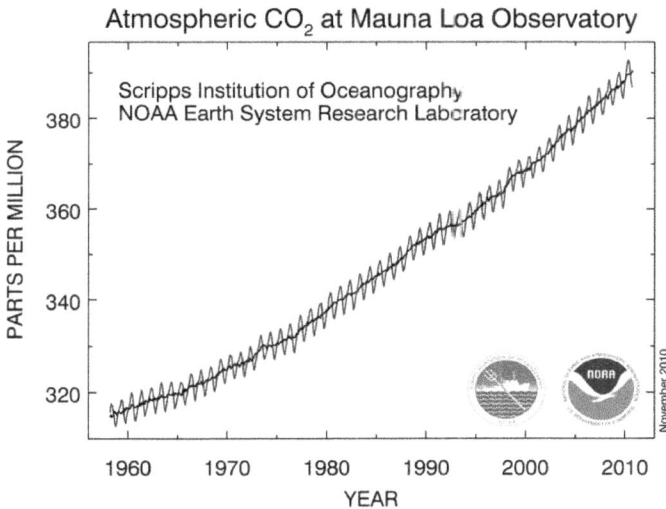

**Figure 36 - This is the graph used universally to show the growth in $CO_2$ and proof of climate change. Source: www.esrl.noaa.gov**

Mauna Loa is Hawaiian for Long Mountain. It is a massive volcano that starts from the bottom of the Pacific Ocean and towers 4100 meters above sea level, forming the core with its sister volcanos to make the big island of Hawaii.

It has very gentle sloping sides, so it looks more like a big hill than a mountain.

The volcano is still active and its last spectacular eruption occurred in 1950. Since 1832, Mauna Loa has erupted 39 times. Its most recent eruption was in 1984. In 1868, it caused the largest earthquake ever recorded in Hawaii, a magnitude 8.

Yet its Mauna Loa's remoteness that make it ideal for monitoring the atmosphere.

Collecting and determining the composition of the atmosphere is complex. Roughly, the Earth's atmosphere consists of nitrogen, (78.08%), oxygen (20.95%) and argon (0.934%). Argon is an inert gas, coming from a Greek word meaning inactive. Together, these gases total 99.964% of the Earth's atmosphere. The remaining 0.036% of the atmosphere contains gases such as carbon dioxide, neon, helium, methane and krypton.

To put this breakdown into parts per million, the scale used to measure $CO_2$ at Mauna Loa, nitrogen is 780,840ppm, oxygen is 209,460ppm and argon is 9,340ppm while $CO_2$ is 390ppm.

To make things a bit more complex, the atmosphere consists of 1% to 4% water vapour. Measurements of $CO_2$ are taken from a dry atmosphere. Almost all of this moisture is found below the height of about 4000 meters. Plants love the nitrogen and $CO_2$ while we people and animals thrive on the oxygen. Everything needs water.

To measure the particles of the atmosphere requires highly specialised equipment and collection techniques. Measuring any gas at 390 parts per million is complex. A person breath, a passing car or the volcano expels some gas and the sample is compromised.

However, it is clear that our scientific community does a very good job in measuring $CO_2$. In fact, not only can the scientists measure $CO_2$ but they can measure the 3 different isotopes of carbon that constitute $CO_2$ - Carbon 12, 13 and 14. These isotopes point towards the original origin of the carbon.

From these measurements it fair to conclude that the levels of $CO_2$ are increasing and these increases are the result of human activity.

The World Centre for Greenhouse Gases has an online database that allows you to download the measurements of gases collected around the globe.[46]

In the media, the dangerous carbon balloons being released into the atmosphere is one of the most visual images of climate change. Carbon is portrayed as an evil gas that will destroy the planet. Of course, not all $CO_2$ is bad for the environment. The question is how much.

What I found very interesting is how consistent $CO_2$ is in the entire atmosphere. Figure 37 shows these measurements for both Mauna Loa and the South Pole. We are told that the consistent fluctuation of $CO_2$ in Hawaii is from the Northern Hemisphere growing season and the impact of plants on $CO_2$ levels. This is not the case in Antarctica. Yet both measurements show similar levels and increases in $CO_2$.

The starting point of the graph is 280 parts per million - a figure generally seen as the pre-industrial era levels of $CO_2$.

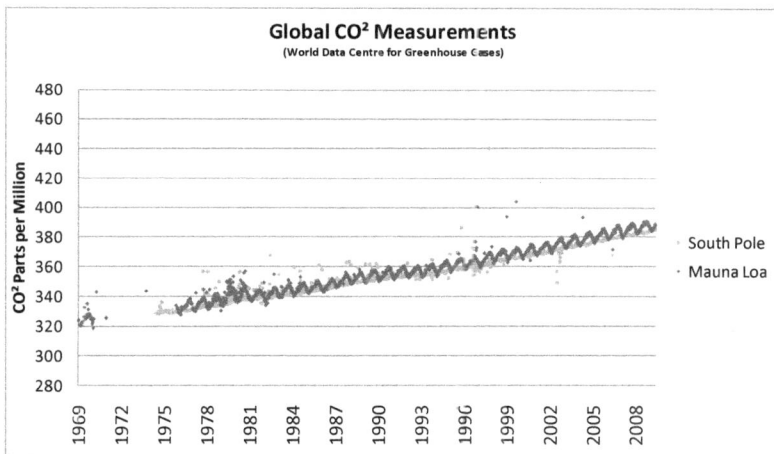

**Figure 37 - $CO_2$ measurements taken at Mauna Loa and the South Pole. The amount of $CO_2$ and its increases in the atmosphere are obvious.**

If you then take the same chart and add more measurement sites, you can see that the levels of $CO_2$ are very consistent around the globe as shown in Figure 38.

The data points outside the norms are a result of sample contamination. These measurements are typically excluded but have been presented here for completeness.

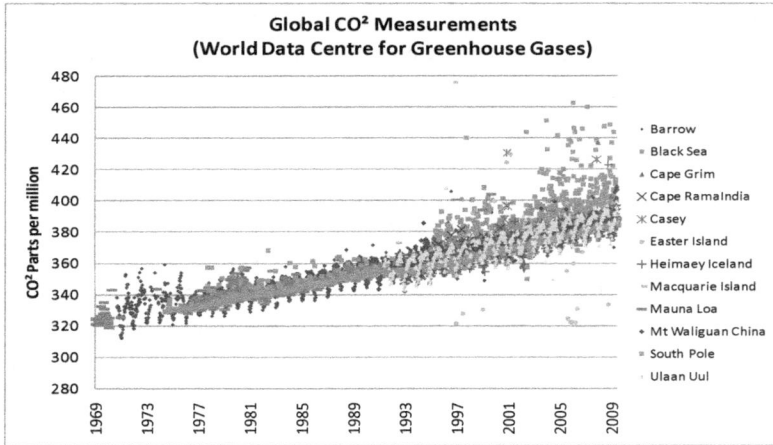

Figure 38 - Global $CO_2$ measurements. Sites include Barrow Alaska, Black Sea, Cape Grim (Australia), Cape Rama (India), Casey (Antarctica), Easter Island, Heimaey (Iceland), Macquarie Island, Mauna Loa Hawaii, Mt Waliguan (China), the South Pole and Ulaan Uul (Mongolia).

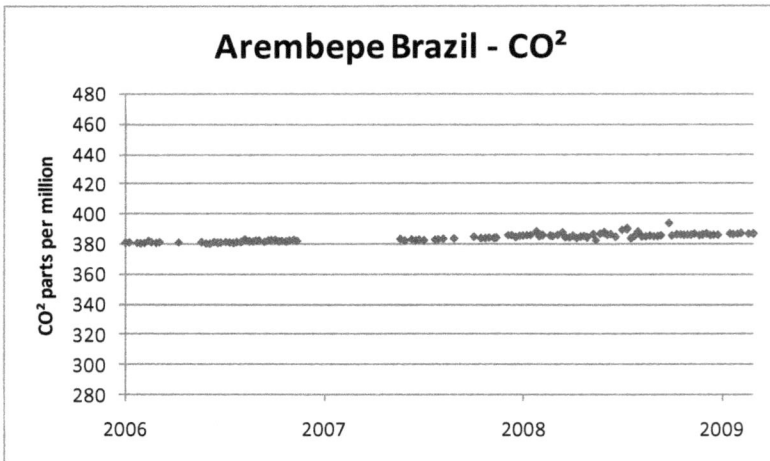

Figure 39 - The limited $CO_2$ data for Arembepe, Brazil shows an increase in levels.

Even if you take the limited data from Arembepe, Brazil, the same trend exists. And while Arembepe is well south of the Amazon, is at least in the general vicinity to be influenced by the rain forest.

$CO_2$ is increasing in all sites using the raw data. However, just because the $CO_2$ levels are rising doesn't mean we will have global warming and climate change.

To assess the impact of increased $CO_2$ on the climate you have to first understand the science behind the greenhouse effect.

Earth's greenhouse effect has nothing to do with a greenhouse to grow plants. It is an unfortunate use of the name. A plant greenhouse uses its glass walls to reduce airflow and warm the air inside from sunlight. With the Earth, heat from the Sun, in the form of solar radiation, hits the earth. About one third of this energy is reflected off the atmosphere back into space. About ½ of the remaining solar radiation is absorbed by the Earth's surface and the planet warms. This warmth is emitted from the Earth's surface back into the atmosphere and some of it back into space. This is where it gets complicated, the greenhouse effect occurs when the wavelength of this energy is such that it is re-directed in all directions by greenhouse gases. By far, the largest greenhouse gas on Earth is water vapour.

The fear is that this is a self intensifying cycle. With human generated $CO_2$, extra heat becomes trapped on the Earth. This warmth brings more moisture, which in turn brings more greenhouse gas in water vapour. This process generates more heat, to a point where the Earth's climate spirals out of control like Venus, viewed as Earth's sister planet, where temperatures are 480°C, hot enough to melt lead and unable to sustain any life.

The IPCC's paper, *What is the Greenhouse Effect*, explains that "in the humid equatorial regions, where there is so much water vapour in the air that the greenhouse effect is very large, adding a small additional amount of $CO_2$ or water vapour has only a small direct impact on downward infrared radiation. However, in the cold, dry polar regions, the effect of a small increase in $CO_2$ or water vapour is much greater." [47]

Just take a moment to understand what the IPCC are saying. There is no impact from $CO_2$ in the tropics; the region where the hurricanes originate. The $CO_2$ affects the Polar Regions. You would also have to assume that there is, albeit a lesser effect, in the temperate regions of the planet. Perhaps the term *Global Warming* was a bit ambitious? Should we have used polar warming or regional warming instead?

Ok, I don't want to get too critical, but we now need to talk about the climate model. You see, the greenhouse theory was formulated based on modelling. I have spent my life writing complex computer models. The one thing that is for certain, a model is written to predict the actual results, which are verified and confirmed by measurements whenever possible. If proper, actual measurements conflict with the model, the model is wrong not the measurements. Sure, if you measure the temperature incorrectly, such as in the sunshine rather than a shaded area, the measurements are incorrect. But provided the measurements are performed well, the measurements are always correct.

In climate models, measurements are often attributed to individual weather events – a cold spell or a heat wave. Modellers shouldn't forget that climates are simply a collection of weather. Don't get me wrong, climate models are very complex. They need to account for an incredible amount of influences, called variables in modelling.

These factors would include surface wind, high level winds, the jet stream, sunshine, historical temperatures, precipitation in its various forms including rain and snow, cloud cover and types of clouds plus thunderstorm activity, barometric pressure, humidity, dew point, seasonal variations, solar output and activity. You also have to include the parameters associated with ocean temperatures, currents and tides as well as coastal terrain and inland features such as mountains, deserts and plains. Monsoons, dry seasons, tropical storms and drought are also required in the model. This just scratches the surface. $CO_2$ levels are just one variable.

With all the resource applied to meteorology for the last century, we are only able to model the weather, you know it as the forecast, with an accuracy of about 2 days in advance and with acceptable risk about 5 days in advance. Environmental scientists are naive to believe their models to predict future climates, years in advance, provide acceptable accuracy.

The CSIRO's Paul Holper and Simon Torok, in their book *Climate Change – What you can do about it*, pointed to the fact that "Australia's most sophisticated climate model, ACCESS, contains approximately 900,000 lines of computer code." Their view was that this level of detail was additional support for the theory of climate change.[48] Not to put too fine a point on it, but 900,000 lines of code is nothing more than a good start. To put this into context, I worked on a model to predict the faults in the manufacturing of prescription eyeglasses. They wanted to predict scratches, wrong prescriptions, bent frames - what they called faults. It used over 10 million scan points and had well over a million lines of code; just to look at the faults in the manufacturing of eye glasses in a factory less than the size of a football field. Think about how many variables and analysis you need to cover the entire planet?

Models evolve over time and improve in accuracy because the data from actual measurements is used to refine the interaction between the variables. There is nothing better for a modeller than to get actual data to confirm the assumptions used in the model.

Modelling is not unique to climate systems. It is a well advanced technique with a body of substantial study and research. Models are often used in astronomy, to analyse complex things like black holes and the big bang. Recently, *Bayesian Model Averaging*, a technique to review the uncertainty inherent in modelling, has been successfully applied to astronomy models with startling consequences.[49] For example, what were considered known facts, such as the age of the universe and its shape, are now being re-examined. So much so, that a doubling of the accuracy of the universe curvature has lead to estimates that the universe could be 250 times larger than initially modelled through visual observations. While this technique is named after the English mathematician Thomas Bayes, who died in 1761, *Bayesian Model Averaging* had its start in the 1960s and received considerable review from the 1970s onwards. It takes what would be considered a reverse approach to the problem – given the data, how likely is the model to be correct. Given its impact in astronomy, consideration should be made to using this technique in climate modelling.

On this point, there is substantial, high quality scientific debate regarding the impact of $CO_2$ on temperature. But to get the levels of climate change forecast, you also need to have an increase in water vapour. There are now two key variables not one. $CO_2$ needs to increase temperatures and this must to lead to additional rain. Without the additional rain, the global warming claims fall well short of the mark.

A throw-away line is used by the IPCC in AR4 that there it is a well established link between warmth and increased precipitation.

Such a vital component of the global warming theory needs more than just a throw-away line. Will locations such as Australia, which are forecasting drought from global warming, and as such will have less water vapour, experience less warming? Why would the tropics, which are impacted very little by additional $CO_2$, as concluded by the IPCC, experience additional rain and warmth?

It's hot in Phoenix but it sure doesn't rain. What is the impact on deserts? Antarctica is a desert. It receives very little new snow each year. Why will it snow more from an increase in $CO_2$?

I apologise for asking so many questions, but the second part of the theory on global warming is not often considered in the media. The focus has only been on $CO_2$.

It is virtually impossible to conduct an actual experiment on this matter, since we are talking about the entire planet and its atmosphere. The scientific method is based on defining a theory, conducting a set of experiments to prove the theory, publishing the work so other scientists can duplicate your experiments and finally, the theory withstands the test of time.

I was watching a classic episode of *Star Trek*, when Mr Spock is being cross examined in the court marshal of Captain Kirk. When asked if Mr Spock witnessed the event in question, Spock replied "if I let go of a hammer on a planet that has a positive gravity, I need not see it fall to know that it has in fact fallen." The law of gravity makes this a scientific fact.

Governments and the media tell us that climate change is a fact – it is not a theory.

Albert Einstein is considered the most brilliant physicist of the last century. He produced *The Theory of General Relativity: E=MC²*. Now if a *theory* is good enough for Einstein why does climate change need to be considered fact when all other science starts out as a theory?

Einstein's *Theory* obviously took unbelievable intellect and analysis. It is also very straightforward to prove. Energy, mass and speed can all be measured. So far, science has not found a condition where $E \neq MC^2$. Yet still they call it a theory. It needs to withstand the test of time. When it finally does, it will become the *Law of Relativity*. Sir Isaac Newton's Law of Gravity was first published in 1687 – it has withstood the test of time.

Climate change is a complicated theory because it involves so many variables. Let's outline it. An increase in $CO^2$ combined with Weather Variables (WV) leads to an increase in temperature, also combined with Weather Variables. This leads to increased rainfall combined with Weather Variables leading to increased temperature (C') combined with weather variables.

$$CC = \uparrow CO^2 \, (WV) \rightarrow \uparrow \,°C \, (WV) \rightarrow \uparrow Rainfall \, (WV) \rightarrow \uparrow \,°C' \, (WV)$$

When I was studying physics at university, I loved the *small number theory*. It allowed you to throw away lots of stuff attached to your analysis because it was irrelevant. In other words, the values were so small they didn't matter.

Unfortunately, weather variables do matter. In fact, they are a larger contribution to temperature than $CC^2$ or rainfall. They are not small and they can't be ignored.

Leading scientists have expressed alternative ideas on how these weather variables and their interaction in the atmosphere impact the global warming theory. Some suggest that the climate models fail to account for the impact of the solar contribution to climate change.[50]  Some have concerns regarding the impact of clouds. There are other issues which deserve discussion ranging from aerosols to volcanic eruptions.

The other thing we have observed is that the Earth's climate has changed considerably in the past.  Ice ages are the most obvious example but periods of warmth and cold, rain and drought, have reshaped almost every part of the planet.  We can see this through fossil records and changes to the landscape.   All this climate change occurred before any man made $CO_2$ came into the picture.

Does the planet Venus really give us an insight into a world of uncontrolled climate change?

While Venus is the closest planet to Earth they are two very different worlds.  They might be approximately the same size, but Venus' atmosphere is 90 times denser than that on Earth.  It is made up of 96.5% $CO_2$ and 3% nitrogen.  This means that the Earth and Venus have approximately the same amount of Nitrogen.

Venus is closer to the Sun, but the extra heat it receives does not account for its 480°C day time temperature. The Russians landed the Venera 9 space probe on Venus on 22 October, 1975.  It took a couple of photographs and confirmed the temperature and pressure and stopped working; killed by the harsh environment.

The differences between Earth and Venus are considerable.

Venus rotates very slowly - a day on Venus is almost a year.  It has no magnetic poles.  Detailed maps from Magellan, the NASA space probe sent to orbit the planet suggest that massive volcanic activity rather than plate tectonics have shaped the surface.   Its dense clouds continually shroud the planet and drop real, toxic acid as rain - sulphuric acid.  These factors shape the climate of Venus. They are not Earth like.

Interestingly, the planet Mars also has an atmosphere of 95% $CO_2$ and 3% nitrogen.  Mars is tilted on its axis like Earth, has seasons and polar ice caps, a magnetic field and a 22 hour day.  While Venus' atmosphere is 90 times denser, pressure on Mars is about 1% to that of the Earth.  Yet even with this thinner atmosphere, Mars has approximately three times more $CO_2$ in its atmosphere than is contained in Earth's.

Like Earth, Mars has also shown evidence of ice ages. This might support the theory that changes to solar output is the cause of cooling and warming, since the sun is a common link between both worlds. Data from NASA's Mars Global Surveyor and Odyssey missions have shown that the ice cap at the South Pole on Mars, consisting mainly as carbon dioxide in its dry ice form, has been diminishing for three summers in a row. The head of space research at St. Petersburg's Pulkovo Astromomical Observatory in Russia, Habibullo Abdussamatov, said that the Mars data suggests that the current global warming on Earth is being caused by changes in our Sun.

As reported in National Geographic News, "The long-term increase in solar irradiance is heating both Earth and Mars," Abdussamatov said.[51] He believes that changes in the sun's heat output can account for almost all the climate changes we see on both planets. Mars and Earth, for instance, have experienced periodic ice ages throughout their histories. "Man-made greenhouse warming has made a small contribution to the warming seen on Earth in recent years, but it cannot compete with the increase in solar irradiance," he said.

In Mars, we have an atmosphere to study that is not affected by human pollution; and it contains $CO_2$ levels in the same ball park as those of Earth. What an excellent tool to measure results and feed into the Earth's climate change models. Yet it has not been done. Mars shows no signs of warming from a greenhouse effect caused by its $CO_2$.

Habibullo Abdussamatov's analysis was not well received. "His views are completely at odds with the mainstream scientific opinion and they contradict the extensive evidence presented in the most recent IPCC," said Colin Wilson, a planetary physicist at England's Oxford University.

The Russians aren't part of the mainstream? Wow, the cold war is really over.

Some might even argue that the Russians know as much about space as NASA. The Americans might have beaten the Russians to the moon, but the Russians did put the first rocket in space, the first man in space, conducted the first space walk, sent the first probe to another planet and landed several unmanned probes on the moon.

**Figure 40 - Polar ice cap on Mars. Recent data from the planet indicate the polar ice cap is receding. Photo: NASA, Dec-05.**

The Russians even built a space shuttle that only budget cuts in a post cold war era stopped. The head of space research at St. Petersburg's Pulkovo Astronomical Observatory, a centre in operation since 1839, should have some clout.

Yet those mainstream researchers at Mauna Loa don't have to go far to see measurements at odds with their mainstream scientific opinion.

Some of the temperatures at Honolulu, Oahu have cooled. Taking GSOD maximum temperatures with the IPCC measurement benchmark from 1961-1990 and comparing it to the distribution from 1991-2010, there has been an increase in days at 29°C and 33°C against a decrease in days at 31°C.

## Honolulu Maximum Temperature Distribution

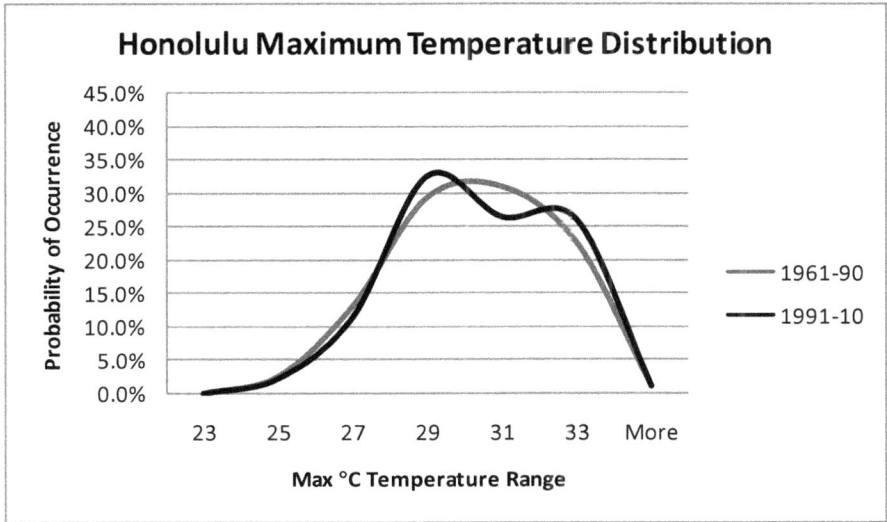

**Figure 41 - Temperatures in Honolulu have cooled, replacing days at 31C with days at 29C.**

Actual measurements show us that the ice cap on Mars has been diminishing for three summers in a row. I'd be adding some more lines of computer code to that climate model.

# Chapter 9
## LET'S GO TO THE SOURCE –
## PITTSBURGH

Back when I was a kid, about 10, I found a coal seam in the woods behind the house. I used to collect the pieces of coal and put them in my wagon. I called it Matic's Mine. Coal was everywhere. My brother and I would crawl into a cave we thought was the entrance to a disused mine. We got so dirty. It seems that Pittsburgh had it all - coal and iron ore, with rivers to transport the steel – and it was a great place to play as a child.

Cut off from the supply of British goods after the War of 1812, Pittsburgh emerged as the steel city. By 1815, it was producing substantial amounts of iron, brass, tin and glass. In the 1830's, Welsh immigrants and their experience in mills and mining flooded the town. By 1857, over 1000 factories were consuming 22 million bushels of coal yearly.

Thousands of working class migrants from Eastern Europe were attracted to the city with the lure of employment.

The city saw rapid growth in iron and glass production during the American Civil War and steel production was extended after 1875 when Henry Bessemer invented the process for the mass production of steel. By 1911, Pittsburgh produced about half of all the steel in the United States, which continued to grow dramatically during World War I. During World War II, the city produced a massive 95 million tons of steel.

Little to no attention was given to pollution control.

The smog and pollution were so bad that street lights had to be turned on during the day. Writer James Parton called Pittsburgh "hell with the lid off" and that was in 1870. He was from Philadelphia anyway, so we didn't care that much.

**Figure 42 – This photograph was taken at 9:20am in downtown Pittsburgh in 1946. Source: Carnegie Library of Pittsburgh.**

Pittsburghers take great pride in our steel town and our football team – The Pittsburgh Steelers. With the highest concentration of steel making in the world, it was Pittsburgh steel that built America. It was Pittsburgh steel that won both World Wars. And it was Pittsburgh steel that polluted the planet.

So with all this pollution, tons and tons of $CO_2$, Pittsburgh should be the warmest place on the planet. Ok, that might be a bit extreme, but you would expect Pittsburgh to experience some effects of climate change?

According to the Union of Concerned Scientists, global warming has already changed Pennsylvania's climate, warning that the high temperatures will affect everything from health to farming.[52] Dr Lewis Ziska from the Department of Agriculture said that cows produce the best in temperatures ranging between 40 to 70 degrees (F), "temperatures above that will result in a decline in dairy production."

Dr Jerry Melillo, director of the Ecosystems Centre at the Marine Biological Lab in Woods Hole Massachusetts, emphasised that skiing and other winter sports may be a thing of the past. "With respect to snowmobiling we think that would be something in people's memory, a historical event, because snow on the ground in Pennsylvania would last a very short time," he said.

**Figure 43 - From the mid-1800s the smoke filled air was so bad that the city suffocated for days beneath darkened skies. Pittsburgh passed the first smoke control ordinance, but World War II delayed its enforcement until the late 1940s. Photo credit: Archive Service Centre, University of Pittsburgh.**

The *Pittsburgh Post Gazette* reported in 2006 that summers in Pittsburgh could resemble those in Georgia or Alabama. "The higher temperatures will mean less snow throughout the Northeast. By the end of the century, the length of the winter snow season could be cut in half and the character of the seasons will change significantly if emissions are not reduced."[53]

The winter of 2010 didn't fit this theory however. On 8 January 2010, ABC News reported a "Hard Freeze from Canada to the Gulf of Mexico – you have to feel for people from Pittsburgh. It has snowed there on every day since December 27."[54]

But climate change is not about one cold winter or one hot day. It's about increasing temperatures in a climate system caused by the greenhouse effects of $CO_2$.

Here the temperature records in Pittsburgh tell a very different story.

The climate in Pittsburgh has not changed. That is if you look at the raw temperature data based on records dating back to 1945.[55]

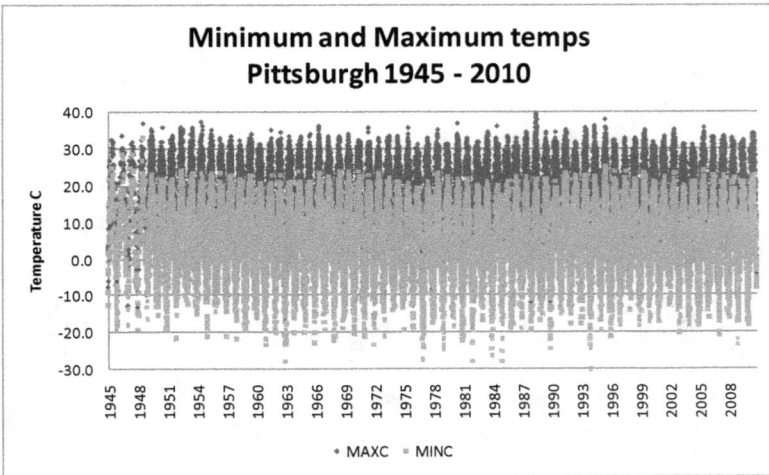

**Figure 44 - The minimum and maximum temperatures (C) for Pittsburgh from 1945. It clearly shows no obvious change in high or low temperatures for the past 65 years.**

The climate rhythm in Figure 44 shows that Pittsburgh has remained very consistent in temperature based on the GSOD since 1945. Comparing the statistical distributions for the maximum temperatures from the years 1945-1979 against the years 1980-2010, Figure 45, shows almost a perfect fit between the curves. The mills began to close in the 1980s. The analysis of minimum temperatures for the same date ranges again shows an almost perfect fit between the two curves. This time there is a slight cooling at 20°C.

The high levels of $CO_2$ in Pittsburgh did nothing to change its climate. The climate also didn't change from the removal of the $CO_2$. Today the mills are almost all gone. Steelmaking has moved to Korea and China. Pittsburgh is a very clean city. The railway lines that once carried the coal, built adjacent to the rivers, are now bicycle paths – rails to trails. Employment has moved from the factory to high technology, with healthcare and computers leading the way.

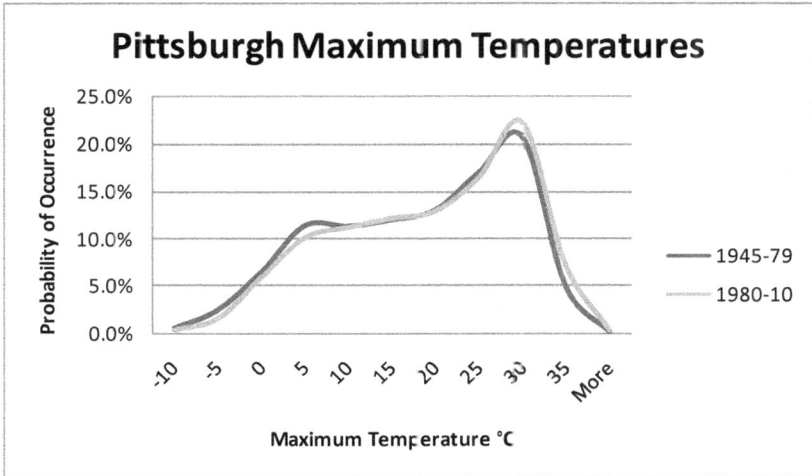

**Figure 45** - The statistical distribution of the maximum temperatures from 1945 to 1979 and from 1980 to 2010. The statistical consistency between the two distributions is clear.

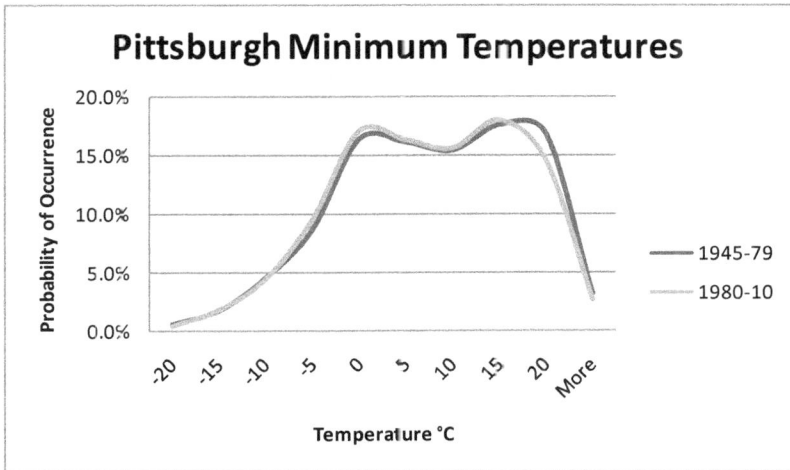

**Figure 46** - The distribution of minimum temperatures in Pittsburgh show the climate is virtually unchanged.

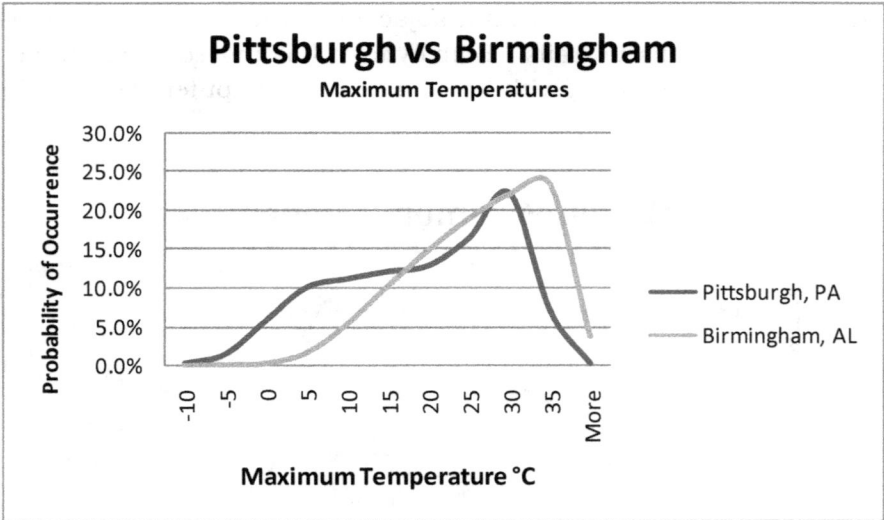

Figure 47 - The distribution of maximum temperatures in Birmingham, Alabama and Pittsburgh, Pennsylvania show considerable differences.

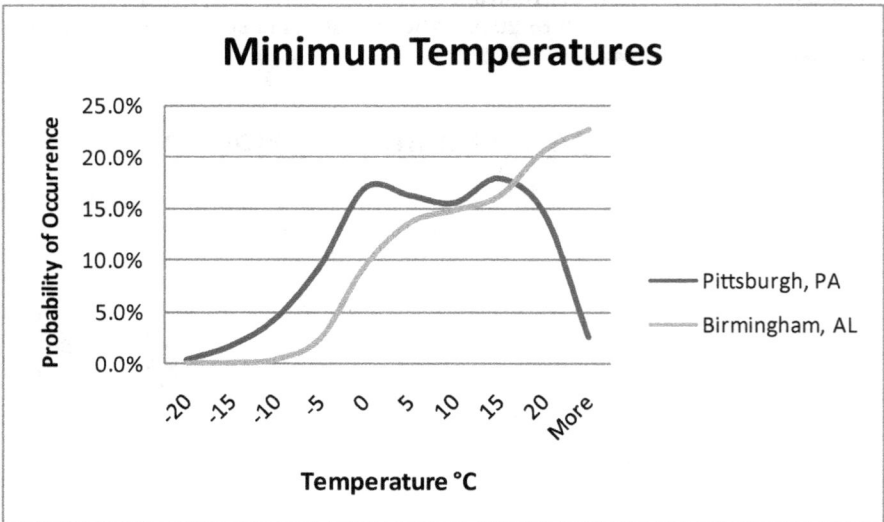

Figure 48 - The climate based on minimum temperatures between Pittsburgh and Birmingham are not even close.

Good things have come from Pittsburgh.

In 1953, Dr. Jonas Slak, a researcher from the University of Pittsburgh, discovered the vaccine for polio. The Presbyterian University Hospital performed the first heart, liver and kidney transplant in 1989. Yet Pittsburgh has influenced almost every corner of globe, it was home of the McDonald's Big Mac.

Pittsburgh is a lot of things, but Alabama :: is not.

If you examine the climate of Alabama and Pittsburgh from 1980-2010, Figure 47, you will see that there are considerable differences in the cold weather and hot weather in Alabama.

But it is the minimum temperatures that really separate the two climates. It just doesn't get cold in Alabama. Pittsburgh with a climate like Birmingham Alabama, certainly not. It is interesting to note that global warming hasn't changed the climate in Birmingham either. See Figure 49 below.

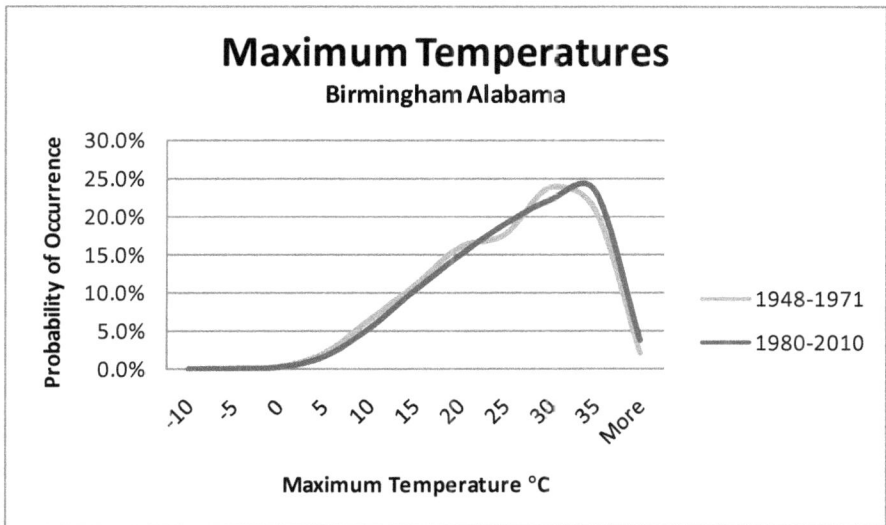

**Maximum Temperatures**
**Birmingham Alabama**

Figure 49 - Global warming also hasn't changed the climate of Birmingham Alabama. Although the records are not as complete as Pittsburgh, Birmingham continues to be Sweet Home Alabama.

$CO_2$, the polio cure, multiple organ transplants and the Big Mac – The world will forever be influenced by Pittsburgh.

# Chapter 10
## MT BLANC

My brother Mike lost his battle with cancer at too early an age. He was 44. He was always a child at heart, so it was fitting that he was diagnosed with *ALL*, typically a childhood form of leukaemia. Mike liked to say that his body never grew up. He was forever young.

He lived life to its fullest. Mike was the most beautiful alpine skier you would ever see. His 6'4" body would just float down a hill, with a smooth swan like movement. He could ski down anything. Steeps, moguls, ice and off cliffs, he never broke away from his graceful style.

In the summer he spent his time at Ohio Pile, a really idyllic spot on the Youghiogheny River, jumping the falls in his kayak. And don't forget the countless trips to Seneca Rocks, West Virginia; the headquarters for rock climbing in the Northeast. We would sleep out in Buck Harper's pavilion. Buck was classic. In every way, he fit the mould of a West Virginia Mountaineer. I'm told that Buck died of a heart attack – impossible anything could kill Buck.

With the winding drive down from Pittsburgh, we always arrived at Buck's Pavilion late at night. At dawn, as the sun peaked through the trees, Buck would be there, kicking you with his oversized boot, and saying "At er be two bucks." You had to pay for the privilege to sleep on a cold concrete floor of a lunch pavilion before you went off to rock climb. "I trust no one," Buck would say. Good advice.

Up on the rocks, our favourite pitch was up the Gendarme, a pinnacle between the north and south faces. But the entire Gendarme fell to the ground on 22 October 1987. October 22 was also Mike's birthday.

It's not often that changes to such a major geologic feature happen in your lifetime. Seneca is made from solid quartzite rock. The Gendarme was documented by the first settlers to the region in 1746 and it was used to train the US troops in assault climbing for action in Italy during World War II. The Gendarme died, Buck died and Mike died.

**Figure 50 - Left, author on top of the Gendarme. Little did I know that I would be one of the last people to climb this unique pinnacle. For those interested, it was an exhilarating 5.4 lead. Right, my brother Mike during our climb up Mt Blanc. Photo: Michael Matic, John Matic.**

Climbers take the thought of global warming very hard. This is a passion that involves being one on one with nature. It is so strange for me to climb in a gym with plastic holds. Rock, dirt and roots are really what climbing is about. It's a mental game against a hunk of granite in the summer or a frozen waterfall in the winter.

So when Al Gore used the summit photos of Mt Kilimanjaro, Tanzania, showing glacier retreat, in his climate change documentary and presentation, the mountain became the poster child of global warming. The images in his documentary are very powerful. Carbon dioxide levels skyrocketing, temperatures souring and ice melting. Climbing will be changed forever.

And if you stop right here, these images make climate change really look like the devil in hiding. But I kept researching.

The complication I found is that the temperature on the summit of the 5892 meter Mt Kilimanjaro does not rise above freezing. For global warming to melt the ice you need the warming part. It has to rise above 0°C and stay there for a prolonged period of time.

Dr Georg Kaser, a glaciologist at the University of Innsbruck in Austria, has been researching the glaciers on Mt Kilimanjaro for some time. He climbs the mountain several times a year. It seems that these glaciers formed about 11,000 years ago, when East Africa was a much wetter place. "Even before the first Europeans reached the summit in 1889, the weather had become dry in Eastern Africa. There simply has not been enough snowfall to keep up with the loss of ice due to sublimation,"[56] Kaser said.

As discussed about Antarctica, sublimation is the same process that causes freezer burn by sucking out the moisture. It is the process of transferring from a solid state to a gas without turning into a liquid in between.

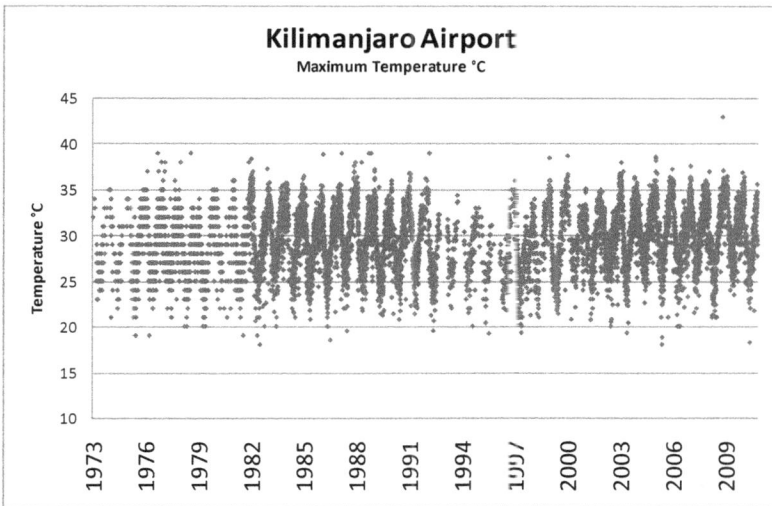

**Figure 51 - The GSOD maximum temperature daily data for Kilimanjaro Airport shows a very consistent range of temperatures since 1973. Please note that this dataset contains a great deal of missing data, something all too common in our climate records.**

"Kilimanjaro is a grossly overused mis-example of the effects of climate change," said University of Washington climate scientist Dr Philip Mote. [57]

He adds, "There simply isn't the magnitude in temperature increases from any type of climate change to cause a major glacier like Kilimanjaro to melt."

Aside from the mountain, if you have a look at the temperature records for Kilimanjaro Airport located near the base of this mountain, you again find no obvious signs of warming. Global warming should also warm the base of the mountain as well as the summit. This dataset, from 1973, has a great deal of missing data, but looking at the GSOD daily maximum temperature records, the consistency of temps at the airport is clear.

Back in 1983, armed with our limited Pennsylvania ice climbing experience, my brother and I went off to climb Mount Blanc, Europe's tallest peak at 4810 meters, departing from Chamonix in France. The memories of this climb are still fresh in my mind some 27 years later.

With the Cold War now over, Mt Elbrus in the Caucasus of Russia is now considered the highest peak in Europe, at 5642 meters. But when we climbed Mt Blanc, it was Europe's highest.

I've not been back to Chamonix since our adventure, but like all of us, I've heard reports that the European glaciers are being ravaged by the effects of global warming - what would Mt Blanc look like today and are my fading slides all that is left of the once glorious ice covered vistas?

Twenty seven years later - our poor preparation, limited food, freezing toes in those leather boots (you really have to thank Bob Lange for inventing plastic ski and climbing boots) and a total inability to read any French in the useless and heavy guide book we took along - I still can't believe we made it up and back down.

But more surprising was a photo I found on the internet, taken by another climber, circa 2008, from virtually the identical location as my original Kodak colour slide in 1983. Refer to Figure 52.

In comparing the two images, the amount of common features are unbelievable. Identical rock lines, profiles and snow lines suggest very little has changed over the years. Even the glacier ridge line is common. The peak in the foreground is not nearly as high as Mt Blanc in the rear, and at this altitude, snow melt and refreezing would be commonplace over the summer.

The photos were clearly taken at a different time of the day, for the bottom photo has substantially more shadowing. However, if the region was subject to major glacier thinning, it is logical to expect more exposed rock and a different rock profile should be evident.

If the upper reaches of the mountain have been untouched by the effects of time, the lower glacier that runs into the Chamonix Valley has experienced shrinkage. The photo on the top in Figure 53 was taken in 1983 as compared to the photograph on the bottom, found on the internet dated June 2006.[58] The glacier in each photo was traced, outlined and overlaid in photoshop, noting the difference with the black outline. It is clearly smaller in 2006.

I needed to access the temperature data to see if these changes were the caused by global warming.

The Grand Saint Bernard Pass sits on the ridge between Mt Blanc and Monte Rosa, the second highest mountain in the Alps, where the remains of the original Roman road still exists. The French artist, Jacques-Louis David, painted what might be the most memorable image of Napoleon in 1800. In his work *Bonaparte Crossing the Alps*, Napoleon sits atop of his rearing white stallion as he directs the 40,000 French troops who snuck into Italy and defeated the Austrian's in the Battle of Montebello.

Yet the reality of the crossing might have been better captured 48 years later by Hippolyte Delaroche's painting of the same name, where the French leader sits looking cold and tired on the back of a mule. These images in Figure 54 were very powerful marketing messages for their day.

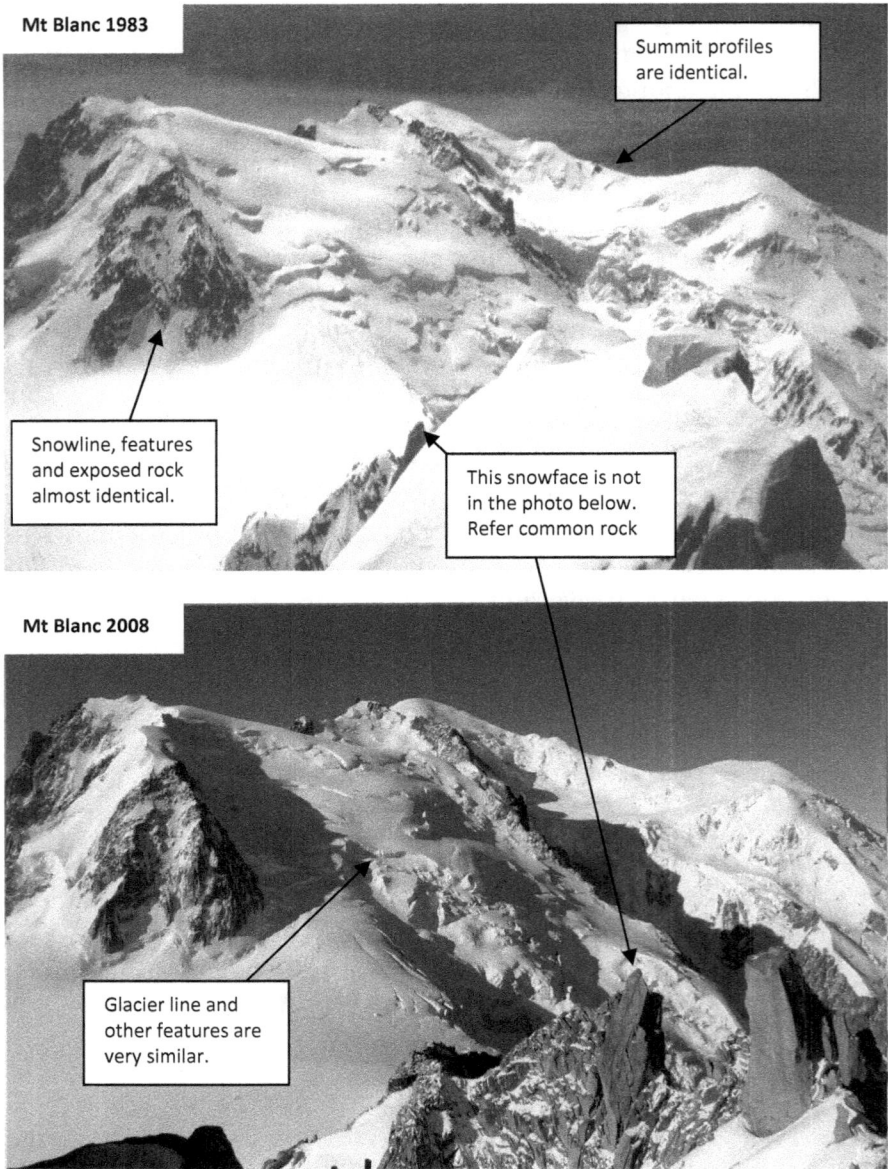

Figure 52 - Photos of Mt Blanc show an almost unchanged mountain over the past 25 years. Snowlines, exposed rock, profiles and general features have not been impacted by global warming. Photo: top – John Matic, Bottom – www.all-free-photos.com.

# Lower Chamonix Glaicer 1983 and 2006

This photograph was taken in 1983 by the author. It shows the lower reaches of the Mt Blanc glacier heading into Chamonix France.

The lower glacier was traced in both photographs and cut out using photoshop.

The lower reaches of the glacier have changed over the years as shown in this overlay. While the colour has been removed for printing, it is clear that the glacier in 2006 is much dirtier than in 1983. Could this factor increase heat and melting at the lower elevations?

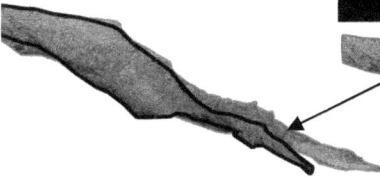

This photograph was taken circa 2006 and found on the web without photographer credit.

**Figure 53 - Chamonix Glacier, at the foot of Mt Blanc, has changed significantly over the years. Of most interest, the glacier today is smaller and much dirtier. Apologies, but the printing of this book required black and white images; the colour version of this image makes the dirt very obvious. This could indicate a relationship between local pollution and glacier melting. A product of urbanisation rather than global warming.**

At 2469 meters, the maximum height of the pass, sits the St Bernard Hospice, which dates back prior to 1050. Built by Bernard of Menthol, he spent his life converting the peoples of the Alps away from their Pagan views. Pope Pius XI confirmed Bernard as the patron saint of the Alps in 1923.

**Figure 54 - *Bonaparte Crossing the Alps* - Two paintings of the same event show very different images. On the left, Jacques-Louis David painted this at the time to symbolise the heroic nature of the event yet 48 years later, Hippolyte Delaroche attempted to capture the event as it really happened, to promote Napoleon as a man of the people.**

Now-a-days, the pass has been made obsolete by a tunnel opened in 1964 and the St Bernard mountain rescue dog, bred on the site as early as 1690, has been replaced by radio beacons. Yet the site did manage to record some very important temperatures in the GHCN database.

The temperature records for Grand St Bernard are very unusual. The database contains every day from 1864-1900, with the exception of 1879, where only 242 days were recorded. The records then stop until late 1964, where everyday but one was measured until 2006. The sample then only averages 284 days a year between 2007 and 2010.

These daily temperatures are plotted from 1983-2006, the time frame in the climbing photographs. As can be clearly seen in Figure 55, the high temperatures are very consistent over the period.

# Grand St Bernard Pass, Switzerland

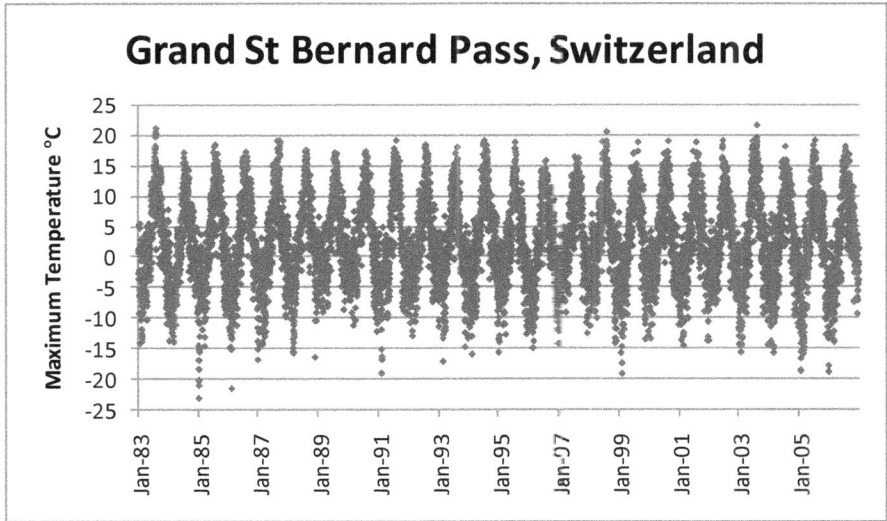

Figure 55 - The daily high temperatures at the Grand St Bernard Pass, on the ridge of Mt Blanc, shows no obvious signs of warming that would melt the Alps' glaciers. This data was used to form the image on the back cover of the book.

# Grand St Bernard, Switzerland

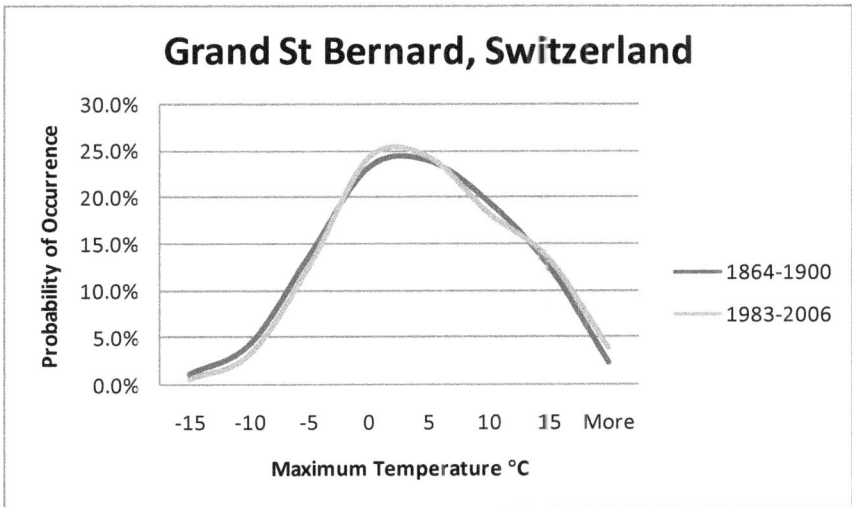

Figure 56 – This analysis of Mt Blanc, measured at the Grand St Bernard Hospice on the ridge line of the mountain at 2469m, is virtually a perfect fit to the previous 142 years of temperature records.

But with this data we are able to perform some very important analysis, comparing 1983-06 with 1864-1900. Europe certainly has changed in the past 142 years, but not the climate on Mt Blanc.

This data clearly shows that warming has not caused the changes evident in the lower glacier. Figure 56 clearly shows that there has been no climate change in this alpine region. I was astonished when this data emerged. These measurements are clearly at odds with the IPCC claims of global warming and the climate model based on grid derived *anomalies*. Even if this example is isolated to one region of the Alps, in all my research, reading countless climate papers and reports, this information has never emerged.

As recently as 2010, Anja Ramming, lead author in a paper on the impact of climate change on alpine plant growth, found that, "the global temperature is increasing and in Earth's alpine regions signs of climate change can already be observed visually by the impressive melting of the alpine glaciers or upward migration of plant species." Using the IPCC warming forecasts, he found that the dynamics of alpine ecosystems will vary considerably by the end of the century from earlier snow melt and a longer growing season. The following extract is how the low resolution temperature data was determined:

> For the future projections, we used gridded data of mean monthly temperature for Switzerland, which were extracted from a dataset for Europe at 10×10 min spatial resolution (~16 km) from 1901–2100 [and] constructed the historical data for the period from 1901 to 2000 from interpolated observations over Europe. They derived future projections for 2001 to 2100 from simulations by two Atmosphere-Ocean coupled Global Circulation Models (AOGCMs). [They] superimposed the AOGCM-derived climate anomalies onto their interpolated observation data, implying that present-day biases in the AOGCM simulation were removed.[59]

I guess using *derived climate anomalies* sounds more impressive than driving up the hill to read that thermometer at St Bernard's Hospice.

The photographic evidence does show one major difference.

The photograph taken in 1983 shows a much cleaner and whiter glacier than the photograph of 2006. I apologise that this is difficult to see in the black and white photo used for book printing. But please take my word for it, the dirt is obvious. As we are all aware, darker colours absorb more heat than lighter colours. In the study of glaciers, this pollution is called cryoconite.

As such, a dirty glacier would attract more warmth and melt faster than a clean glacier. So what we have is photographic evidence showing no impact to the higher reaches of the mountain over the past 27 years; no obvious temperature differences in the past 142 years; yet glacier thinning and increased dirt accumulation at the lower elevations.

Could the problem simply be the pollution from increased road traffic through the Mount Blanc tunnel rather than a global climate change? Either way, Human's are responsible.

A Financial Times Harris Poll, conducted in the United States and the five largest European countries, found in 2009 that "Americans under 65 are less likely than Europeans to see climate change as a major threat."[60]

Certainly several European locations show evidence of a changing climate. For example, I found a decrease in warmer days with more hot days in Vienna and Sonnblick, Austria. Little change was observed in the Swiss alpine town of Montana, yet just 270 km away in the Alps near Liechtenstein, change was observed in Säntis, Switzerland. But finding quality raw temperature data for Europe was a common problem.

This brings us to Brussels, the capital of new Europe. The suburb of Uccle is home of the Royal Meteorological Institute of Belgium. In keeping with this dual language country, in French it's the L'Institut royal météorologique de Belgique and in Dutch, the Koninklijk Meteorologisch Instituut.

GHCN records for Uccle are very impressive. They start in 1833. The only trouble - the records are almost perfect for 166 years, missing only one day; however from 1999 to 2010, they only average 303 days per year. So using the GSOD, I developed a full data set by combining the two temperature databases.

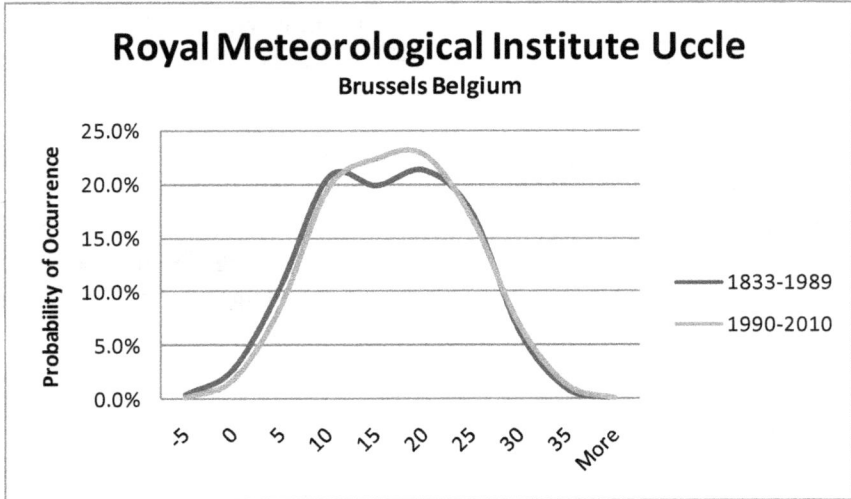

**Figure 57 - The temperature in Brussels was constructed with two overlapping datasets of GSOD and GHCN datasets to create the 177 years of data. Again, it was surprising that no major change in climate is observed.**

The overlapping temperatures from 1973–1999 again show a problem with the world's datasets. In the 9082 overlapping records, only 3.2% of these were the identical value. The largest difference was -12.6°C and the average difference was 0.49°C across the sample. Building statistical models on monthly average grid *anomalies* is certain to fail if the raw data is not valid.

Given this limitation, the 177 years of data, the longest record analysed in this book, shows that the climate in Brussels is virtually unchanged. The capital of New Europe is safe.

# Chapter 11
# THE ARCTIC

**Figure 58 - Map of the Arctic showing the key temperature analysis sites.**

The tragic demise of the polar bear was identified early in the climate change impact assessment. It is regarded as the first animal that will become extinct from climate change.

I've only ever seen a polar bear in the zoo, unless you count the stuffed one at the ABOM Restaurant at the Mt Buller ski resort in Australia. Who doesn't love a polar bear? No one would wish to see them become extinct.

I guess that's everyone who doesn't have to live near them. At up to 680 kilograms and 3 meters high, they are double the size of a tiger and are easily our largest land predator. Ice bears, as the Norwegian's call them, are also highly intelligent and play an important spiritual role for the Arctic people; when they are not being hunted. The Canadian Government Renewable Wildlife Economic Development Office will sell you a non-resident hunting license for $50 and a Polar Bear Trophy Fee for $750, plus 7% GST. I'm not sure how renewable a dead bear is. The polar bear returns the favour however. They have been known to stalk humans for days on the ice pack before the kill. Yet attacks are rare, but that might just be because not many people live in the Arctic. Wildlife officers in central Nunavut, Canada culled several bears in 2010 when they were spotted only two blocks from the elementary school.[61]

As early as 2002, NewScientist magazine reported that the planet's 22,000 bears were under threat based on a report from the World Wildlife Fund (WWF).[62] Lynn Rosentrater, co-author, said "as sea ice is being reduced in the Arctic, the polar bear's basis for survival is being threatened. The sea ice is melting earlier in the spring, which is sending the polar bears to land earlier, without them having developed enough fat reserves."

National Geographic reports went one step further. They claimed in 2007 that new US Government studies found that two-thirds of the world's polar bear population will be gone by 2050. Kassie Siegel a climate change activist with the Center for Biological Diversity in Joshua Tree, California, said, "the report represents a watershed moment in the climate crisis. If we don't change the path that we're on now, then it will be too late, polar bears will become extinct."[63]

Similar to how the British Antarctic Survey led the studies of climate change on the Antarctic Peninsula, the Arctic focus is based around the *Arctic Climate Impact Assessment (ACIA)*. This report was commissioned by the Arctic Council, consisting of Government members from Canada, Denmark, Finland, Iceland, Norway, Russia, Sweden and the USA. The amount of work in this report is impressive. The analysis goes well beyond examining warming in the region and makes assessment on the lifestyle, economics and impact to wildlife in the four identified world arctic zones.

The ACIA was published in 2004 and uses the Intergovernmental Panel on Climate Change emissions scenarios combined with the increases to annual mean temperatures to conclude that the "increased atmospheric concentrations of green house gases are very likely to have a larger effect on climate in the Arctic than anywhere else on the globe;"[64] summarised in this extract from the report:

> Over the Arctic, the ACIA-designated models project a larger mean temperature increase: for the region north of 60º N, both emissions scenarios result in a 2.5 ºC increase by the mid-21st century. By the end of the 21st century, Arctic temperature increases are projected to be 7 ºC and 5 ºC for the A2 and B2 scenarios, respectively, compared to the present climate.

These conclusions were, and are still, widely reported in the media. Warming in the Arctic is twice that of the rest of the planet.

I was waiting in my doctor's office when I saw a National Geographic Magazine sitting on the table. The issue, June 2010, has a photo of a cute hut, covered in green moss, with the headline:

# *Greenland*

## *Ground Zero for Global Warming*

The story reports that "satellite measurements show that its vast ice sheet, which holds nearly 7 percent of the world's fresh water, is shrinking by about 50 cubic miles each year. The melting ice accelerates the warming—newly exposed ocean and land absorb sunlight that the ice used to reflect into space. If all of Greenland's ice melts in the centuries ahead, sea level will rise by 24 feet, inundating coastlines around the planet."[65]

I started reading the article in depth. I was certain that with such a headline combined with the quality journalism and photographic resources of National Geographic, the evidence of climate change would be overwhelming. But the story and the headline didn't equate at all. It seems that the people in Greenland are still waiting for their land to 'green' since they were lured to the island a millennium ago by Erik the Red, a Viking from Iceland. According to his *Saga*, he called it Greenland to attract more migrants with a favourable name. Today it would be called a real estate scam.

The National Geographic story is about Greenlander's waiting to see whether the "greening of Greenland, so regularly announced in the international media, is actually going to happen."

So back to the GSOD and GHCN temperature databases.

Lonely Pituffik, Greenland, is home of the US Thule Air Force Base. It is a really good site for talking polar temperatures, but I'm not sure if Colonel Mark E. Allen and his command of about 550 residents drew the short straw. Home of the 821st Air Base Group, it provides defence and support for the Ballistic Missile Early Warning System designed to detect and track ICBMs launched against North America. I guess news that the cold war is over has yet to reach Pituffik.

Yet the US Air Force puts on a brave face to encourage recruits: "Thule's arctic environment offers some of the most spectacular scenery found anywhere in the world, including majestic icebergs in the North Star Bay, the massive polar ice cap and Wolstenholme Fjord - the only place on earth where three active glaciers join together."

Combine this with the fact that Thule is home of the northernmost deep water port in the world, makes this zone a climate study Eden.

The GSOD contains temperatures from 1952 – almost 60 years of records. It does somehow forget the years of 1971 and 1972, but other than that, the collection is virtually complete.

So has the climate at Thule changed since the Korean War? Is it warming at twice the planet's rate, in support of the IPCC greenhouse gas models?

The answer would have to be no. Again, remember where we are talking about. Ground zero for global warming – so where's the warming?

You could put a real critical eye on the chart in Figure 59 and say that there is a slight difference in the maximum temperature distribution between the 13,508 days in the sample from 1952-1990 and the 7281 days from 1991-2010. Is this difference material?

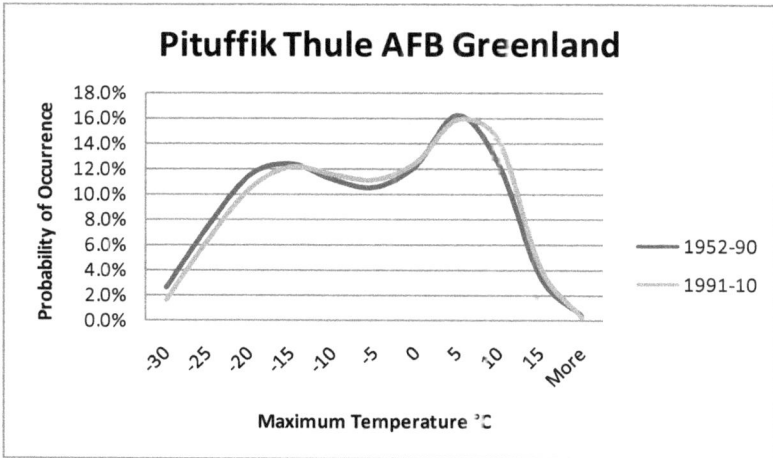

**Figure 59** - The maximum temperatures for Thule Air Force Base in Pituffik Greenland show very similar climates between the years 1952-90 and 1991-10. There is no suggestion that the temperatures are twice the global average. Note that 1971 and 1972 are not contained in the sample.

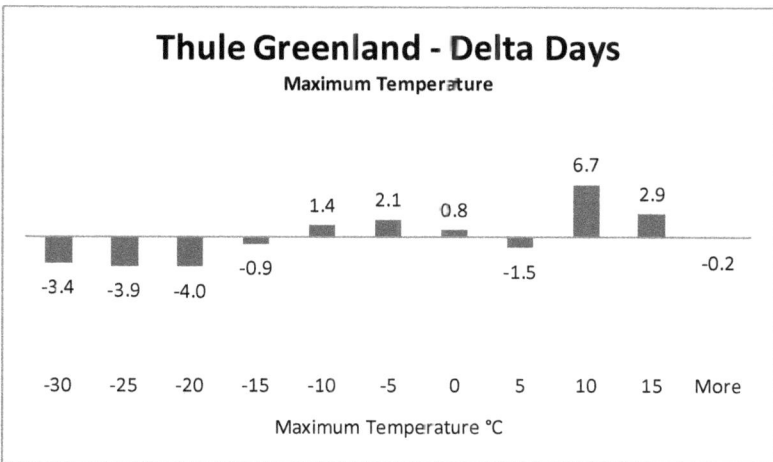

**Figure 60** - The variance between the two climate distributions allow the delta in days per year to be calculated.

## Thule Greenland
### Original Climate and Delta Days

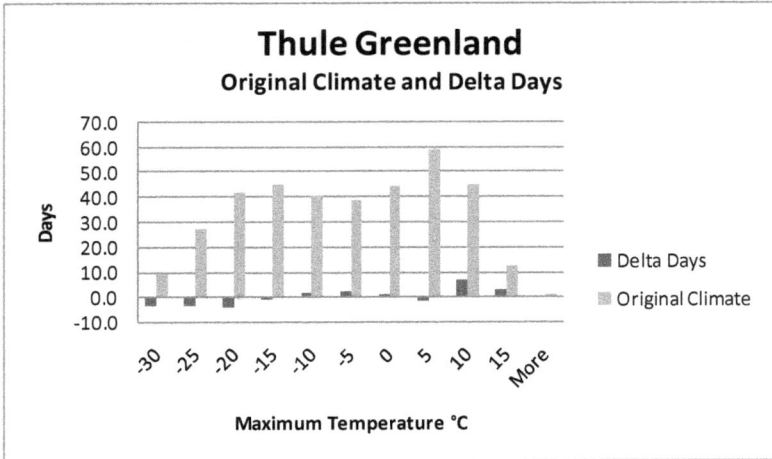

**Figure 61 - Comparing the original climate with the delta days shows a very slight difference in the two climates.**

## Pituffik Thule AFB Greenland
### Minimum Temperature

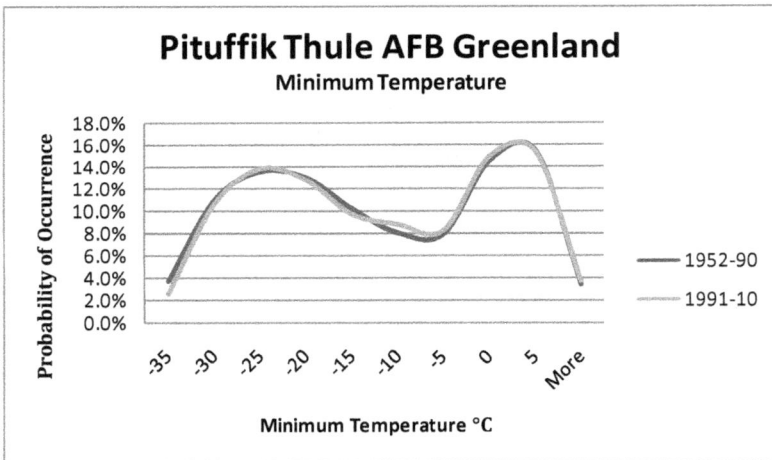

**Figure 62 - The minimum temperatures at Pituffik Thule Air Force Base in Greenland show almost identical profiles and no impact from climate change.**

Now comes the fun part. Taking the variance between the two distributions allows us to calculate the difference in days, called *delta days,* for the temperature ranges.

So, Thule averaged 4 less days per year when the maximum temperature reached -20°C and 6.7 more days when the temperature reached 10°C. (Figure 60)

The final piece of analysis is applying these delta days to the original climate, in this case from 1952-1990. There were 45 days in the year at 10°C, so will an extra 6.7 days be enough to melt the ice that is up to 3 kilometres thick?

Oh, before I forget, this is only half of the analysis. The same calculation occurs for the minimum temperature. The good news is the temperature distribution is so close, that if is fair to conclude that there is no material difference in the minimum temperature of the climates.

In summary, the lows are identical and the highs brought 8 additional days a year, out of 118, where the temperature was above freezing. There is no global warming at Pituffik.

The 3396 people that live in Qaqotog, located on the southernmost tip of the island, also have trouble recording the temperature. They did record 99.6% of the days from 1974 to 1990, but only 97.5% from 1991 – 2010, with most of the days missing being recent, since 2004.

Based on this, they don't appear to be experiencing global warming either.

The missing days for Qaqotog are obvious in the daily temperature records, but even with these gone, there are no obvious signs of significant warming. Remember, Greenland is supposed to be warming at an alarming rate – at least twice that of the rest of the planet.

The two remaining Greenland sites in the temperature database are Tasiilaq, located on the Southeast coast and Danmarkshavn, a weather station in the Northeast, with a population of 8 people. It must be a very lonely place to work.

These two sites are important to analyse the east coast of Greenland. A region influenced by the turbulent Denmark Straight and its ocean currents.

## GSOD Temperature Records - Qaqortoq

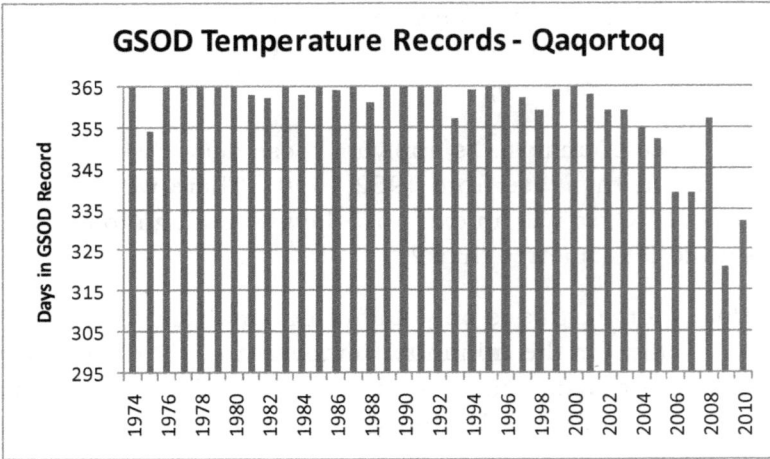

Figure 63 –Missing temperature records for Qaqortoq show that recent missing days make comparisons with previous years difficult. This is a problem all too common in the climate record.

## Qaqortoq

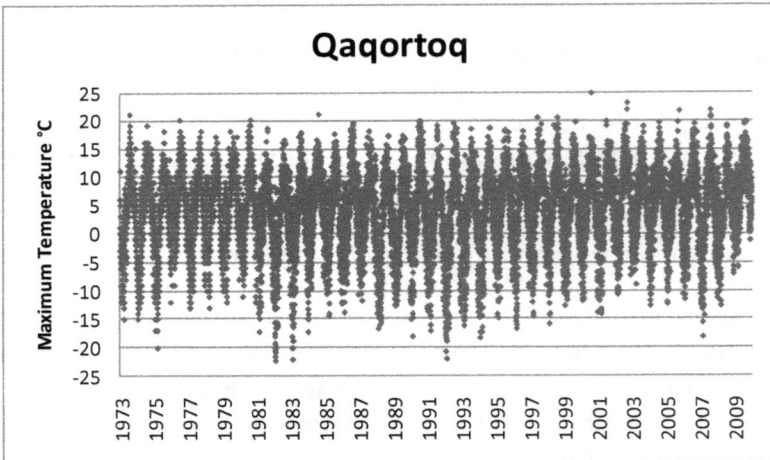

Figure 64 - The daily temperature records for Qaqortoq Greenland. The missing temperature records are obvious in the later years. Even with this missing data, no obvious signs of significant warming are evident.

Unfortunately, Tasiilaq is missing on average 100 days in each year since 1973. Not one year is complete for this important location. The same is true for Danmarkshavn. In Figure 65 and Figure 66, the turbulent autumn month of September is graphed for each location and compared to 2009. No significant difference was noticed.

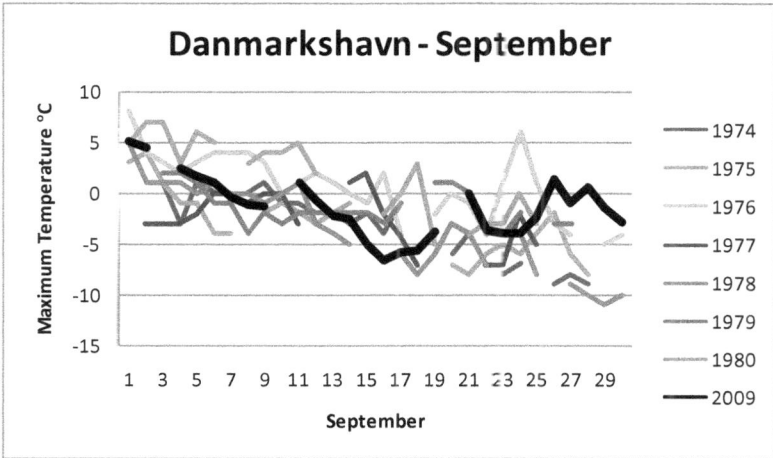

**Figure 65 – The Danmarkshavn weather station, a very important location, has not one full year of temperature measurements. The measured days are shown in black for September 2009 against 1974-80. A typical profile for September in 2009 was found.**

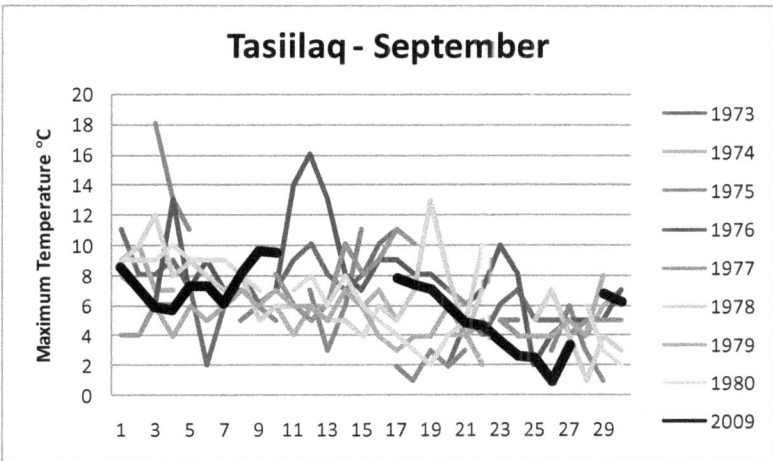

**Figure 66 - While Tasiilaq is a very important site, there is not one full year of temperature measurements in the temperature database. This analysis for the month of September shows that 2009, the thick black line, was in-line with a typical profile during 1973-80.**

Der Spiegel, one of Germany's leading publications, reported in 2006 that global warming is a boom for Greenland's farmers.

The magazine suggests that the average temperature in Qaqortoq has increased from 0.63°C to 1.93°C in the last 30 years.[66] Yet local agronomist Kenneth Høegh cautions this bounty, saying "potatoes do grow in Greenland these days, but not so very many just yet."[67]

Such an important location in the climate change debate and Greenland has only one location with reasonable records for analysis in the GSOD or GHCN temperature database.

Leaving Greenland and moving further southwest, is Canada's Newfoundland. Not technically in the Arctic, the station on Stable Island has been taking temperature records since 1897.

Visually, the daily maximum temperatures on Stable Island are very stable. (Ok, a bad pun.) The temperature distribution from 1915 to 1990 is constructed and shown in Figure 68. The next step is to overlay the data from 98 to 14. No, this is not a typo. That is 1898 to 1914. So if you've been critical on the curves not fitting exactly, have a look at Figure 69. There is a slight speed hump at 5°C.

Climates will vary somewhat over years and decades. What the distributions show is a pattern of data that represents the overall climate. What really matters to the ecosystem is when there are major changes in the extremes.

The IPCC climate model (Figure 6) forecasts this. It claims we should expect to see less very cold days, more hot days and more record hot days. If you now overlay the final graph, the years from 1991 to 2010, you see that Stable Island shows no signs at all of any changes to either extreme element of the distribution. (Figure 70) As with 1898-14, there are a few strange things happening between 5°C and 13°C in 1991-10.

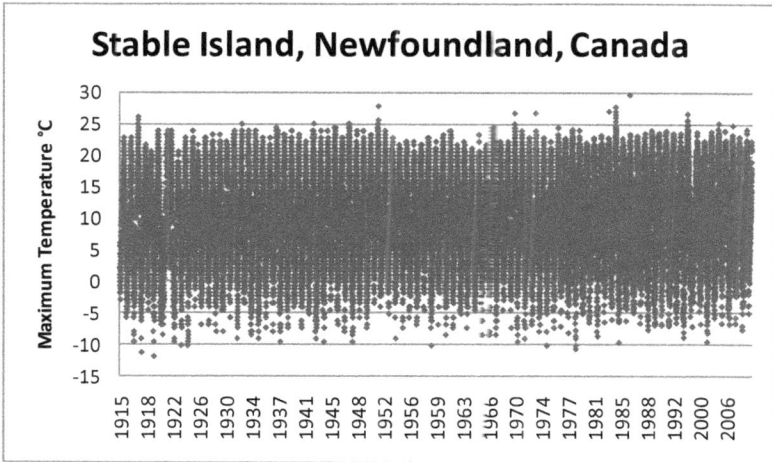

**Figure 67 – Maximum temperatures for Stable Island, Newfoundland, Canada since 1915. No obvious signs of warming trends are present. Data exists from 1897, but excel would only graph 32,000 data points.**

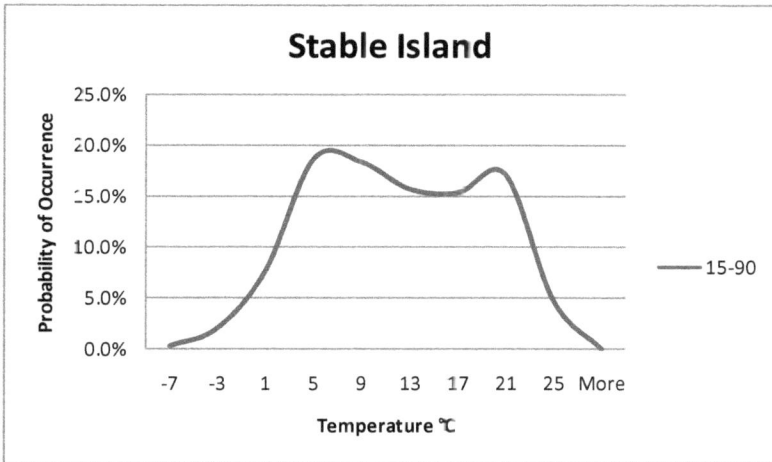

**Figure 68 - The distribution of maximum temperatures for Stable Island for 75 years from 1915 to 1990.**

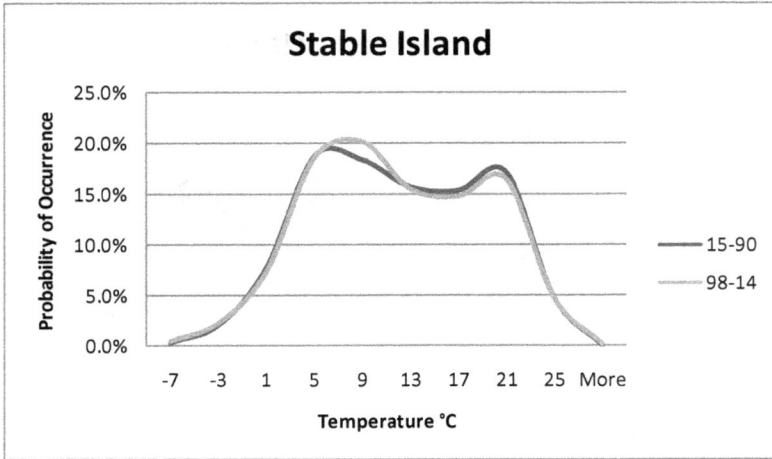

Figure 69 - The second graph is an overlay of temperatures for 16 years from 98 to 14. You might think it shows some effects of climate change around 5C, until you realise that the overlay is from 1898 to 1914.

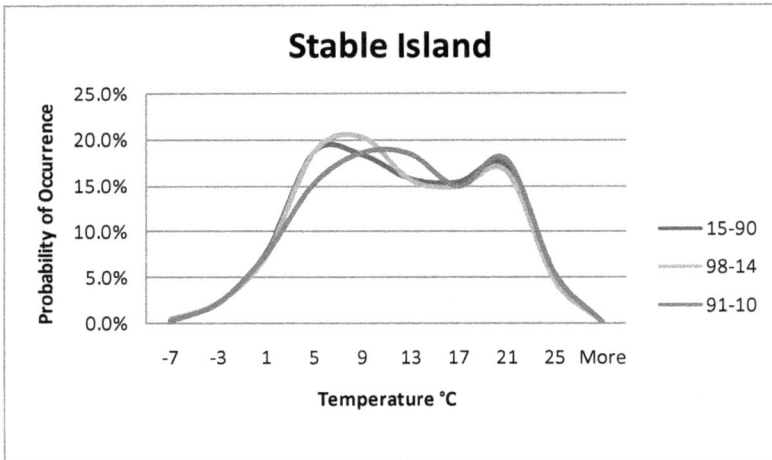

Figure 70 - Distributions of maximum temperature for Stable Island for 1915-1990, 1898-1914 and 1991 to 2010. No changes in climate have been recorded in any of the temperature extremes.

The 229 people of Resolute Canada, called Qausuittuq in the native Inuktiut language, make up what is regarded as the coldest inhabited and northernmost community on Earth.

Named after the HMS Resolute, a ship that became trapped in the ice and abandoned, it was eventually recovered and its timbers used to construct the Resolute Desks, used by the Queen of England and the President of the United States. You can only get so much from history, even if your desk did feature in the blockbuster film *National Treasure – Book of Secrets*. I'm sure this town would welcome a bit of global warming. And in Resolute, I found some. Since 1947, the warmest day in Resolute was 18.5°C in 2008. But before you throw away that Canadian goose down parka and celebrate climate change, there has actually been no change in the distribution of warm days for Resolute. While the town has experienced fewer very cold days from 1991-10, as compared to 1947-90, it has traded these off for more cool days rather than more warm days.

South of Resolute, across the Queen Maud Gulf, is the town of Cambridge Bay. Here temperature records exist from 1949. Climate change at Cambridge Bay has brought less extreme cold days but more cold days. There is minimal change to the warm day profile (Figure 72).

This data leads us to the obvious question on the plight of the polar bear and the amount of sea ice available for it to hunt seals. Does such a change in the climate have impact on the sea ice?

To answer this, let us start by examining what a winter looks like in Cambridge Bay. There does not seem to be any question that adequate sea ice existed in the 1950s. Of note, a polar bear's life span in the wild is approximately 25 years, so we are talking about two generations, and multiple years of breading for the of bears.

Winter starts early in Cambridge Bay. It drops below freezing from early October and by the middle of the month, it is cold. But don't be fooled that the temperature is consistent, a conclusion that is easy to draw from the smooth temperature distributions. The days are very volatile.

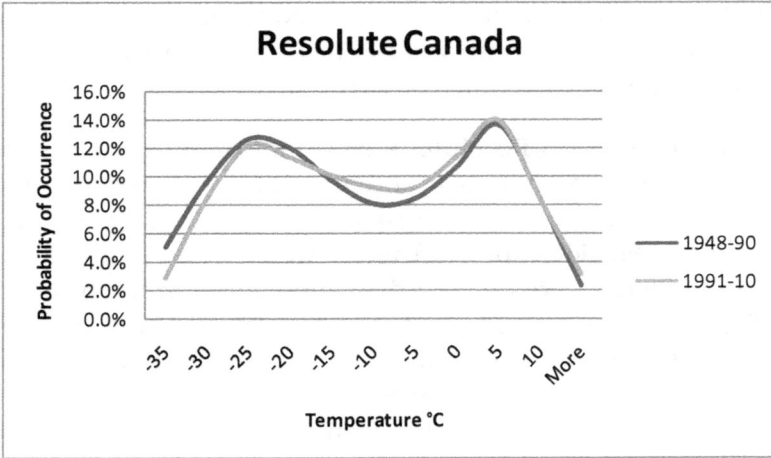

**Figure 71 - Resolute Canada has traded 2.1% of its days at -35C for 1.2% more days at 5C, but global warming has failed to give Resolute the desired increase in warm days.**

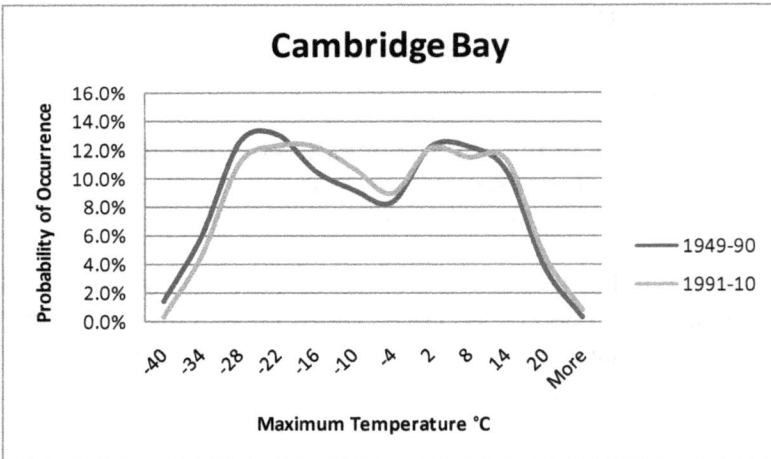

**Figure 72 - Cambridge Bay Canada is experiencing less very cold days but more cold days. No change to the warm day profile has occurred.**

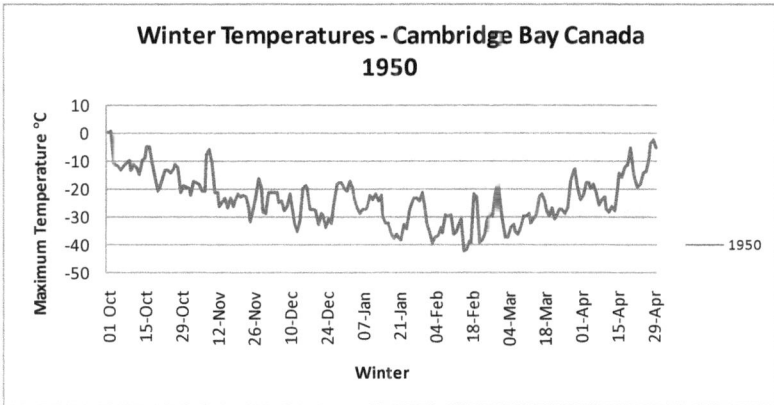

Figure 73 - Winter temperatures in Cambridge Bay fall quickly from October to January, where they gradually increase back to freezing by May.

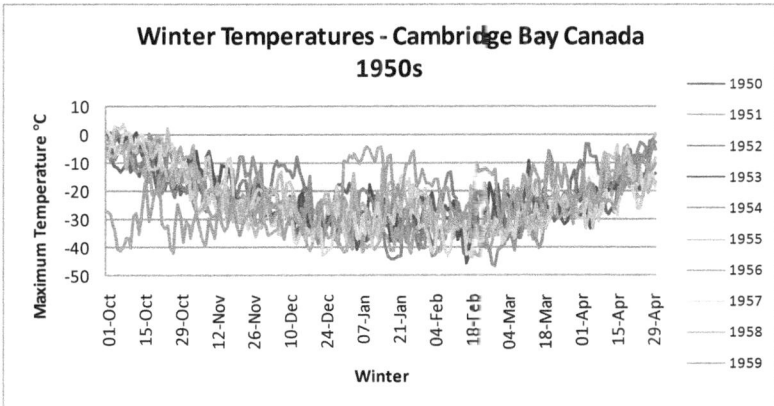

Figure 74 - The decade temperatures all followed a similar trend until 1959, which started out extremely cold but experienced dramatic warmth in January 1960.

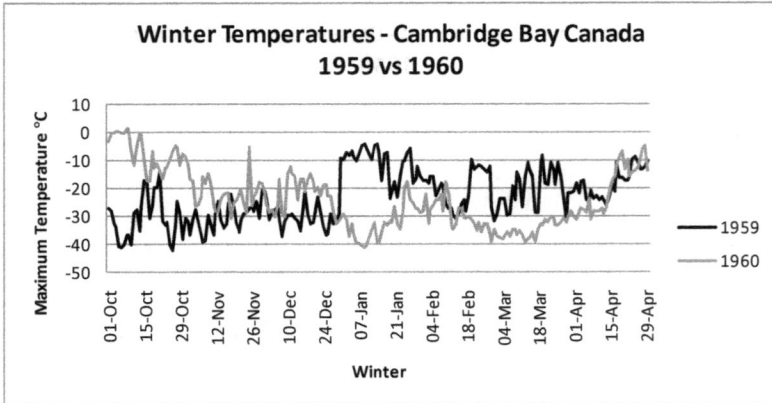

**Figure 75 - The extreme winter of 1959 as compared to winter 1960, a more traditional year.**

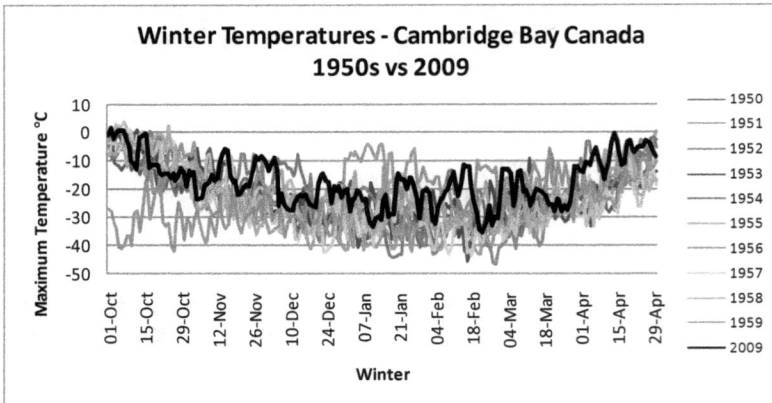

**Figure 76 - Winter 2009 overlaid against the winters during the 1950s. It was a warmer winter, but nothing to suggest that sea ice would not form or thaw in the same manner.**

While I was plotting the 1950s, it all seemed remarkably consistent. Sure, you had movements of -10 to -20 degrees in a day, falling close to -50°C, but the trend seemed very consistent. That is, until I plotted 1959. This year started out very cold, falling to -41.1°C on 6 October. But on the last day of December, the temperature rose 20 degrees, and stayed there throughout January; and never really received any of the extreme cold again that winter. By 1960-61, Cambridge was back to its same old self.

Why 1959 was so unusual is not clear. Winter 2009 was warm by Cambridge Bay standards. Yet the media reported a harsh winter that was widespread from Canada to the Gulf of Mexico.[68] Yet even being a warm winter, 2009 did not feature anything out of the norm. Sure, colder days existed, but so did warmer days. The sea froze in the 1950s and there is nothing in the air temperature to suggest that it wouldn't freeze and thaw in a similar manner in 2010. Refer to Figure 76.

Eureka! If you google map Eureka, Nunavut, Canada, you'll find a small weather station on the Fosheim Peninsula of Ellesmere Island, the large island just to the west of Greenland. Literally, the middle of nowhere.

Eight staff have the honour of operating the station on a rotation basis, down from its peak of 15 in the 1970s. Established on 11 April 1947, the temperature measurements start on 1 May 1947. You have to compliment the people who built and staffed Eureka. It contains almost a complete set of temperature measurements, missing only 46 days in the 22,992 days since 1947 - 99.8% complete.

So we now have the two key components to measure climate change: Temperature measurements for 63 years and a station high in the Arctic.

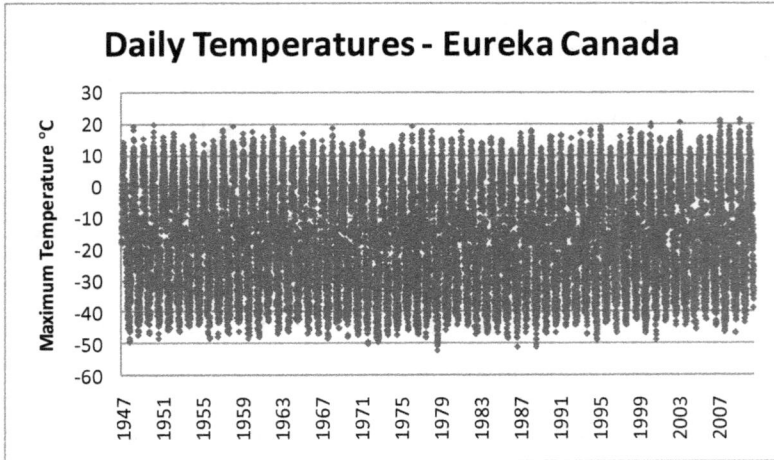

Figure 77 - The daily temperatures for Eureka Canada. High in the Arctic with almost complete measurements, this weather station provides excellent in-site into climate change.

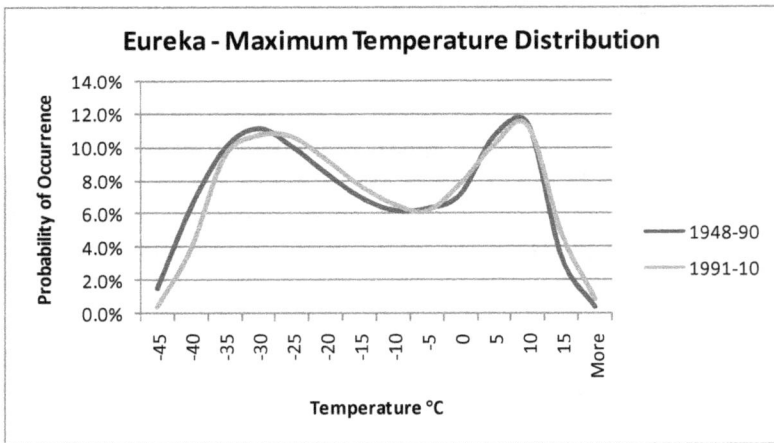

Figure 78 - The maximum temperature distribution for Eureka Canada shows very similar patterns for 1948-90 and 1991-10. It would be difficult to make the claim of significant warming based on the data.

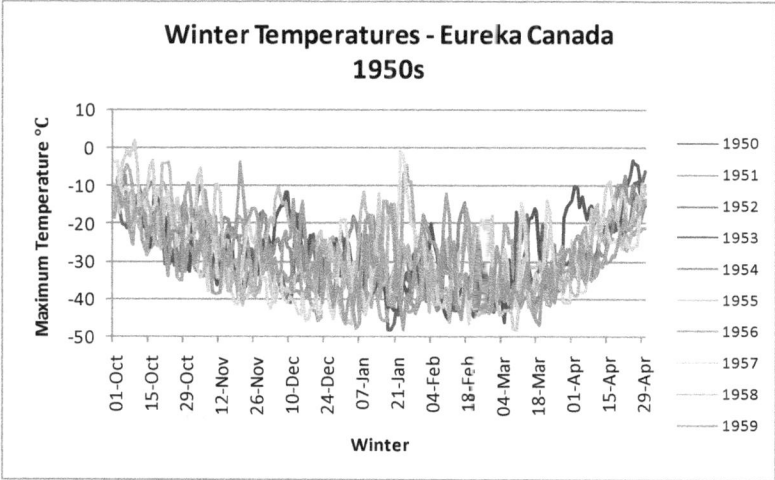

**Figure 79 - Winter in the 1950s at Eureka Canada, high in the Arctic. While very low temperatures are typical, so are volatile changes. Note the warm weather for January in 1957 and 1951.**

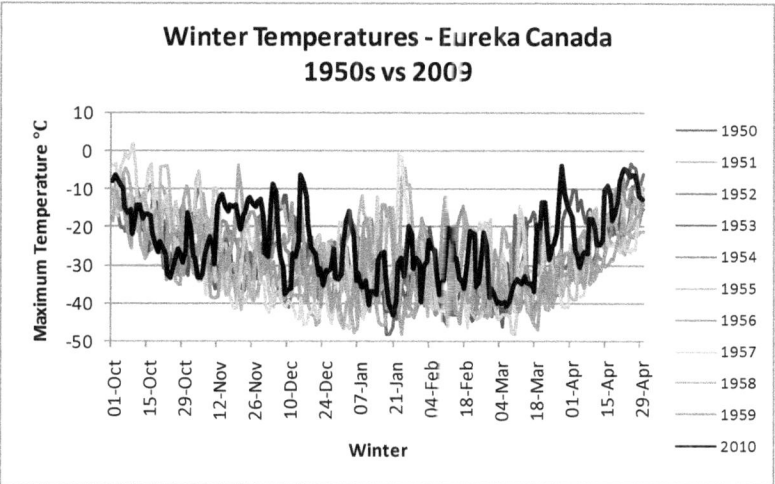

**Figure 80 - Winter 2009-10 started out very cold, but a warm spell in November moved to more typical weather by January. A second warm spell near April was probably responsible for the rain witnessed.**

On 14 July 2009 the temperature at Eureka reached its maximum since measurements were made, 20.9°C. This high temperature and reports of rain in April near the North Pole were reported as evidence in support of global warming.[69]

So has the climate of Eureka warmed since 1947? You don't need one off events to make these claims when you have access to temperature records.

Certainly from the graph of daily maximum temperatures, the climate in Eureka is extreme and volatile.

Like Cambridge Bay (Figure 72) where the distribution shows a trend from extreme cold to cold days, Eureka has a very similar distribution but less variance between the two samples. (Figure 78)

In the depth of the winter, on the 23rd and 24th of January, the temperature at Eureka Canada, high in the Arctic, in total darkness where the sun's rays bring no warmth as they fail to rise above the horizon, and you would expect the maximum temperature to -40°C, the thermometer rose to -1.1°C and -1.7°C before falling back to -11.7°C on 25 January.

But before you declare that the polar ice caps are melting and global warming is upon us, the year was 1957.

In 1951, it was above -10°C for four days in a row, from 24 to 27 January, hitting a high of -4.4°C. Winter in the 1950s was turbulent, as seen in Figure 79.

Yet not all the Arctic is behaving in the same manner. Moving further west to Alaska, the cities of Barrow and Nome tell two different stories.

Barrow shows the presence of moderate climate change in-line with the IPCC model. I'm sure its 3982 citizens would welcome a bit of warming, with 204 days a year below freezing. These changes have brought 11 extra days above freezing.

Barrow receives very little moisture, with only 127mm of equivalent rainfall, mainly falling as snow, per annum. It is completely overcast with low stratus clouds and fog for more than half the year.

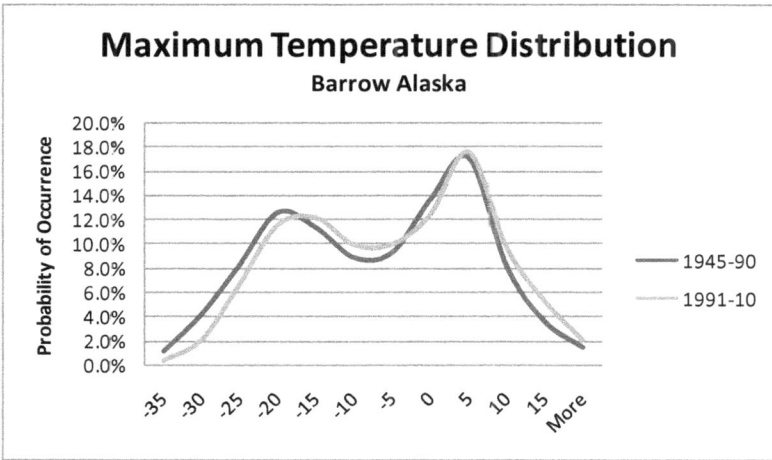

**Figure 81 - Barrow Alaska has had changes to its climate, with less extreme cold combined with more warm days.**

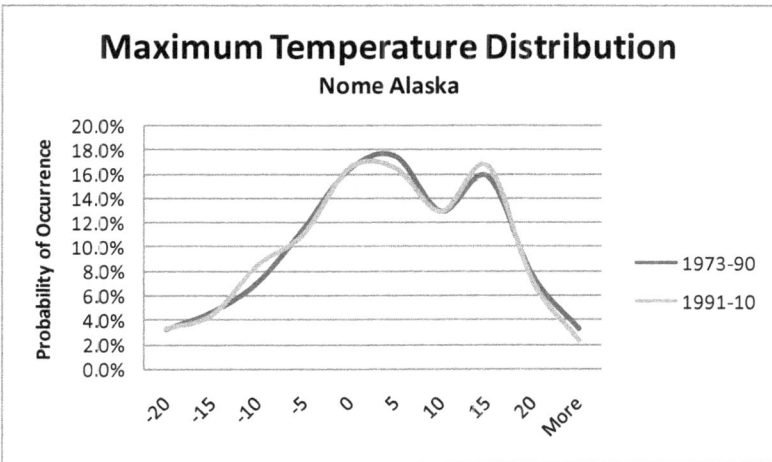

**Figure 82 - Nome Alaska has show some climate cooling, with more cold days and less extreme warm days.**

Sitting with the Arctic Ocean on three sides and 300 km of flat tundra to the south, it is frequently bucketed by strong winds. 3982 brave souls!

*There's no place like Nome.* This is the slogan for the city, so I can't lay claim to it. Nome was incorporated in 1901 and is most famous as the finish for the 1049-mile Iditarod Trail Sled Dog Race from Anchorage each March and the famous sled dog Balto.

In January 1925, diphtheria was raging through the town. Heavy snow and freezing temperatures prohibited the aircraft of the day to make the flight with the life saving serum. So off went the dog sleds to save the day. On 2 February 1925, Gunnar Kaasen drove his tired dog team into Nome, one of the 20 drivers who took part in the 674 mile relay from Nenana to Nome.

Our South Pole hero Roald Amundsen also spent time in Nome, learning the art of dog sledding and adopting the Inuit way of life. These are the skills that allowed him to reach polar success. During his journeys, he noted that in the year 1903, the Pacific was free of ice.

Nome has experienced some cooling, with more cold days at -10°C and less extreme warm days.

Across the Bering Strait we find the country with the greatest Arctic presence, Russia. Again, there is a large problem with the temperature database in regards to missing records.

The Russian's have outstanding, long term Arctic records, in many cases, dating back to the turn of the century. They managed to measure the temperature during the First World War, the Russian Revolution of 1917 and the civil war turmoil that followed into the 1920s, the Great Purge in the 1930s and the Nazi invasion during World War II. However, it is obvious that many of the records in the contested lands were lost. For example, temperatures for Moscow don't start in earnest until 1949.

Think of the Russian weather technician who got the gig to record temperatures in the White Sea Port of Arkhangelsk rather than being set off to defend Stalingrad? Not one day during the war years was missed. They did miss 60 days in 1947. It is not clear exactly why this occurred.

The problem with Russian temperatures is that after the fall of communism, records from 2000 – 2010 are very patchy. Just when we need the data to analyse global warming, the data is not kept.

To digress for the moment, I'd like to take you back to Operation Barbarossa, the Nazi Germany invasion of Russia which commenced on 22 June 1941. History books and documentaries often tell us that the Nazi army was halted by an unusually cold winter on the gates of Moscow. The failure to outfit their army for the winter conditions was the turning point on the front.

I was curious if this claim was in fact supported by the temperatures.

If you take the temperatures during the 1930s in Tobolsk, Siberia, the closest site with measurements during the war and compare these to the war years, you will find that the extreme cold days were very similar. The war years had an increase in days around -5°C with a decrease in days of 5°C.

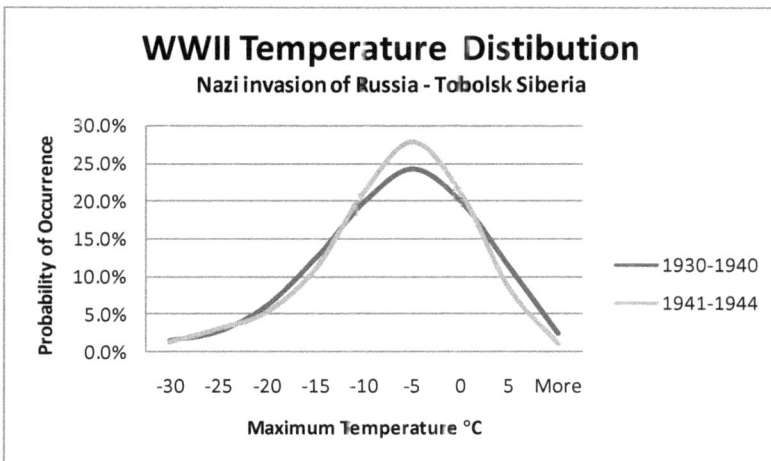

**Figure 83 - A comparison of the maximum temperature during the 1930s and comparing this to the years of World War II. The Nazi's should have known the climate would get very cold.**

Based on these measurements, there was no sensible reason why the German army should not have been issued winter uniforms from the start of the campaign. I found this interesting.

Now back to the present. Just recently, I was watching the documentary *Wild Russia* on television. This episode covered Wrangel Island or Vrangelya in the Russian Arctic. It is located just across the Chukchi Sea, to the west of Barrow and Nome. In the midst of the snow and ice, freezing temperatures, July summer snow and rutting muskoxen, the narrator informs us that this world won't last; it is melting from global warming. Wrangel Island is a breeding ground for polar bears. It has the highest density of dens in the world.

Temperatures for Wrangel Island go back to 1926 when a team of explorers took advantage of the ice free and calm waters and landed on the island with 3 years of supplies. Soon after they landed, the weather turned and the only thing that saved the party was the courage of the captain of an ice breaker who arrived in 1929.

The measurements are near complete, but they stop in 1999. Still, unless something really crazy happened during the last 10 years, the temperatures on Wrangel Island are of little concern to the polar bears. This arctic environment isn't ending any time soon.

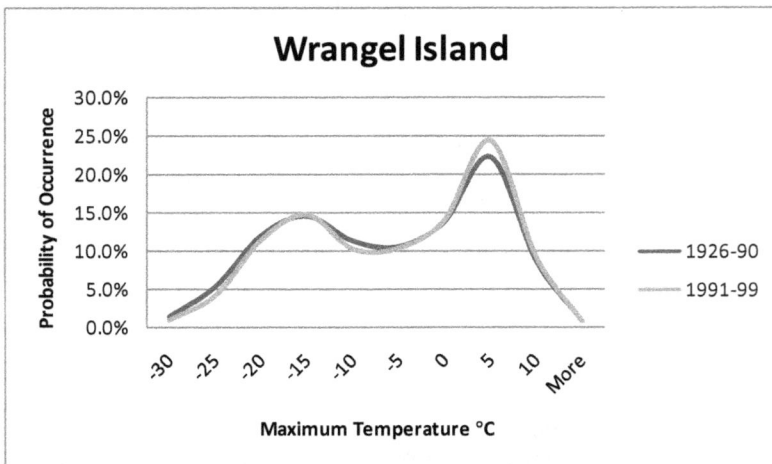

**Figure 84 - Wrangel Island experienced no substantial climate change during the 1990s.**

Still further to the west is Kotelny Island, part of the New Siberian Islands between the East Siberian and Laptev Seas. This should be a good site for climate analysis, but the measurements were only taken for an average of 231 days per year. There is also a great deal of duplicate looking data. For example, the days from 19-22 September 1960, all had the same temperature of -1.1°C. This is common, in examining the dataset, days are often repeated. Yet this is from the GHCN database. It is supposed to be more accurate and receive a higher level of audit. Given the volatile nature of arctic climates, multiple days of identical records should have set the audit bells ringing.

The city of Khatanga, on the Russian mainland, is one of the most northern settlements on the planet. Again, the analysis of its temperature distribution is limited to 1998, as the past 12 years has averaged only 262 days of data. This said, Khatanga is one of the strangest distributions measured. This is probably because the temperatures are very volatile and the comparison of 8 years is not sufficient to form a smooth climate. This said, Khatanga has had more cold days between -30°C and -25°C and more warm days at 5°C. The distribution of extreme warm days is identical.

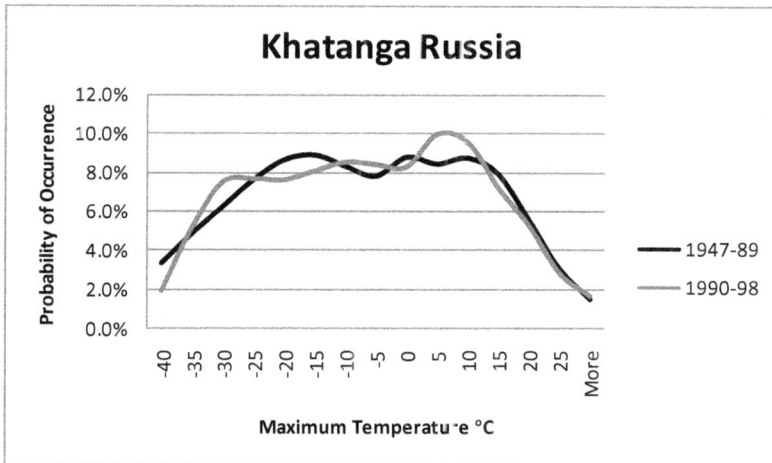

Figure 85 - The maximum temperature distribution for Khatanga contains only 9 years, so this might be why the fit is so poor to the original climate from 1947-89.

It's a very funny distribution. I'm not sure it tells us anything about climate change but rather that climates can be unpredictable.

That said, each Arctic site analysed can be grouped into two similar trends - Fewer extreme cold days but with little change in the warm or hot days; or little difference in the climate. Both of these trends do not suggest that air temperature is melting the sea ice.

Sea ice is always fighting a constant battle to stay frozen. The ice is immersed in the liquid ocean, which has a temperature above freezing and is influenced by ocean currents that travel as far away as the tropics. The tides of the underlying sea also continuously move and fracture the ice. The source of freezing is the air temperature and the ice freezes from the ocean surface downward.

As shown, when analysing the daily temperature profile in the Arctic, some climates have been changing, but not at the alarming rate forecast by IPCC models. But the sea ice in the Arctic region is reducing. NASA claims that its satellites have found that between 2004 and 2008, multi-year ice cover shrank 1.54 million square kilometres - nearly the size of Alaska's land area.[70] Yet not all the ice is vanishing. The first year ice, as NASA calls it, usually with a thickness of about 2 meters is increasing in volume, while the multiyear ice, which averages about 3 meters, is reducing. Overall, NASA's evidence shows that the total ice volume is reducing.

While the satellite data only examines 5 years, declassified data from US submarines, covering approximately 38% of the Arctic Ocean, found the mean overall winter thickness was 3.64m in 1980 as compared to 1.89m during the last year of the satellite measurements. Combining the submarine and satellite analysis shows a long-term trend that spans five decades of sea ice thinning. The submarine data also showed that thinning in the earlier years occurred in a major portion of the perennially ice covered Arctic Ocean.[71]

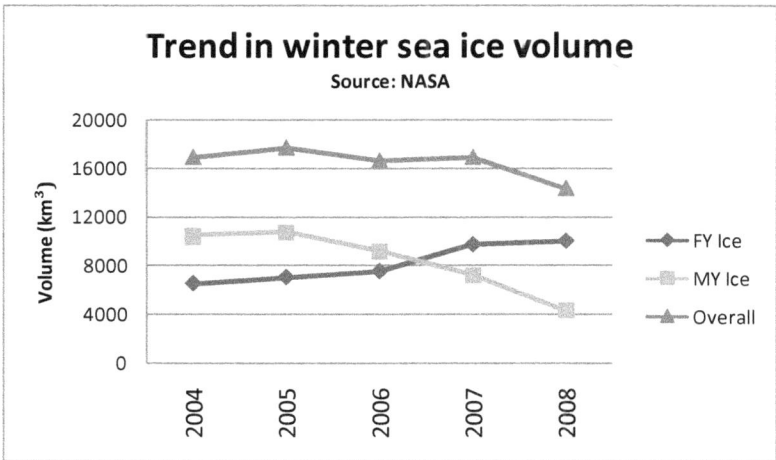

**Figure 86 - Trend in winter sea ice volume. This graph, produced by NASA is available on their website. It shows a decline in multi-year ice (MY Ice) and overall ice but an increase in first year ice (FY Ice).**

The key issue here is that reducing the number of less extreme cold days cannot account for a decrease of 1.75m in thickness of ice covered ocean. Like the Larsen B ice shelf in Antarctica, surface warming is too simplistic a model to account for the Arctic ice thinning. Something else is happening in regards to the currents in the oceans and our research is blinded by incorrect perceptions that the climate in the Polar Regions is significantly warming.

While 60 years of declassified submarine records sounds impressive, if you head back to before the cold war, the Arctic was a major battlefield during World War II. The Russians utilised the northern sea route from the Alaska to Arkhangelsk, its White Sea port, to provide vital military supplies. The Germans attacked the convoys mainly with capital ships, such as the massive pocket battleship the Admiral Scheer. Outgunned, the Russians were successful with their convoys because the Germans assumed they would hug the coast, while the Russians, knowing that the actual summer ice conditions were predictable and not as harsh as believed, allowed them to sail deep into the Kara Sea, to the north of the Novaya Zemlya Islands.[72]

Maybe the Arctic ice is a bit more dynamic then first anticipated?

NASA has also found that sea ice in Antarctica has been advancing northward by about 1% per decade; the equivalent of 100,000 square kilometres.[73] Researchers are at a loss to explain the difference between the Arctic and the Antarctic, but changing ocean dynamics is one of their key theories.

Suitably, we should finish our Arctic odyssey on Bear Island. It is called Bjørnøya in Norwegian – above the Arctic Circle between the Greenland and Barents Seas. Nowadays, the island's only inhabitants are nine staff in the weather station. The island has been declared a nature reserve for the creatures that have survived its exploitation. You see, Bear Island's walrus population was hunted to virtual extinction from 1609. This time is referred to as the *Grand Era* - not so grand if you're a walrus. The island was also utilised for whaling, sealing and fishing. Eggs of seabirds were aggressively harvested until 1971. And funny enough, from 1918 to 1932, the entire island was a privately owned coal mine.

It must be a captivating nature reserve. Very few plants grow on the island, mainly moss. There are no trees. Not one glacier. A few arctic foxes do make a living from feasting on the sea birds. The walrus is rarely seen.

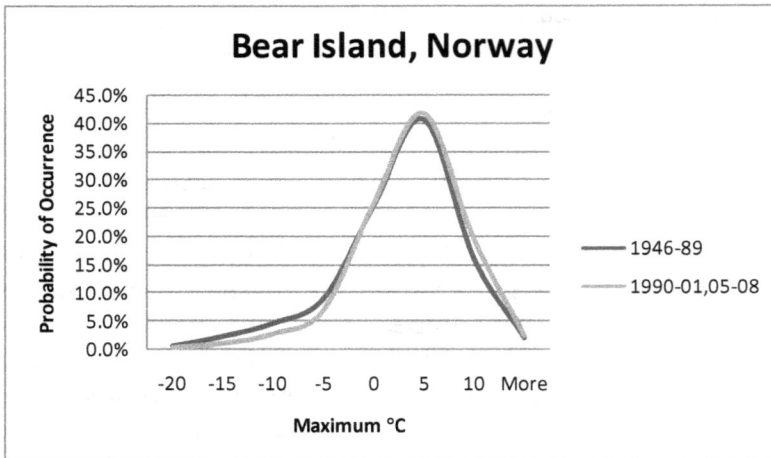

**Figure 87 - Bear Island has a very mild climate for a location so high into the Arctic. Yet it is not able to sustain a human population. This distribution was taken from 1946 to 1989 and compared to 1990 to 1991 and 2005 to 2008. The other years were missing too much data to be included in the comparison.**

The good news, there are still polar bears from whom the island is named. Good thing they didn't call it Walrus Island. The Norwegian Meterologisk Institutt warns us that bears represent a safety problem and often visit the weather station in the winter months. Sometimes they spend the summer on the island as well.

A great deal has changed on Bear Island, just not the climate. Refer to Figure 87.

Polar bears extinct by 2050? My money's on the polar bear.

In Canada, Jonathan Pameolik, a government wildlife technician, offered his comments as to why two nanuq, the local name for polar bears, were shot in Whale Cove and nine bears were killed in Arviat. "The ice flow edge came really close to the community, and I guess that's why they're having bear problem there."

In the paper's online blog, Dal Martell commented: "Nanuq is on top of the food chain, do you really think it will be extinct by 2050? I think you're underestimating the Polar Bear on its' hunting ability. I really think this animal is adaptable and will find other food sources to survive - ie. other land mammals - caribou, which are always overly abundant, wolves, foxes, wolverines and ground squirrels in their dens, molting birds, musk oxes, to name a few."[74]

Wildlife officials in the region are warning people to keep a close eye on their children and people are also advised not to travel outside without carrying a gun. Signal pistols and the rifle are obligatory equipment for weather personal on Bear Island as well.

You've got to watch out for those hungry bears.

# Chapter 12
## AUSTRALIA

Fears of Canadian insurrection, from its French colonies and US sympathies following the American victory in its war of independence, helped England decide it needed a better place to dump its convicts. In many respects, Australia can look to Canada for its modern existence.

Today, both countries love sport. But they seldom play each other.

Australia's ice hockey team did qualify for the 1960 Squaw Valley Winter Olympics; but they failed to win a game, losing to Czechoslovakia 18-1 in their first international. Australia did hijack mogul skier Dale Begg-Smith from Canada and subsequently won what most Canadian's believed was their Olympic Winter Gold Medal.

Both countries also fear climate change.

For over a decade, Australia's climate change policy was shaped by the Howard Government. Some would argue that we had no climate change policy. I always found it funny that the conservative, right wing of politics are called Liberals in Australia and the socialist, left wing of politics are called Liberals in the United States. By the way, a Republican in Australia is someone who wishes to see the Queen of England removed as the head of state. So this term doesn't bridge the divide either.

So the Liberals, under the conservative icon, John Howard, held power in Australia from 1996-2007. He was swept into power with 53.6% of the population giving him their vote. Actually, in Australia you vote for you dislike least. It's kind of like a game of survivor. Each ballot usually has multiple candidates, the least favourite is eliminated and their votes are put back in the count for the next candidate to be eliminated. This continues until there is one standing. It's called preferential voting.

Of course, like most democracies, the voting is done per election district, so you can win office, even by a landslide, with less popular support than your opposition. When the Howard Government was re-elected in 1998, they only received 49% of the vote.

During the Howard era, climate change debate was effectively removed from the National agenda. In February 2006, the Australian ABC investigative television program *Four Corners*, asserted that a powerful lobby, called the *Greenhouse Mafia*, representing Australia's industrial sector, were successful in undermining any attempt to introduce climate policy. Four Corners also reported that CSIRO scientists were gagged from speaking about climate change.

Holding Government with 50% of the vote means that half the country actually doesn't like you. Curtailing the debate forced supporters of climate change to become vocal. Tackling climate change became a policy of the political Labor and Green opposition.

Such an example is Clive Hamilton, chair of the Climate Institute, who gave a widely reported speech to the Australia Institute in 2006. He named the *Greenhouse Mafia* and took the debate to a new level:

> I hope that in 50 years time as Australians swelter in debilitating heatwaves, battle fierce bushfires, fight over dwindling water resources, lament the loss of unique species and tell stories recalling the wonders of the Great Barrier Reef, they will be reminded of the names of those who refused to act in the face of overwhelming evidence of what lay ahead. They carry a huge burden of moral responsibility, and I hope that their descendants will understand the shameful role that they played.

But the politics in Australia changed in 2007 with the election of a Labor Government. Kevin Rudd was swept into office with 52.7% of the vote, what the media called a *Ruddslide*. With his slogan, *Kevin07*, he labelled climate change as "the greatest moral, economic and social challenge of our time." He promised a cut to greenhouse gas emissions of 60% before 2050. His first official act as Prime Minister was to sign the *Kyoto Protocol* – a UN agreement setting binding targets on the reduction in greenhouse gas emissions.

At 73%, Rudd enjoyed the highest approval rating of any Australian Prime Minister.  Yet behind the walls of power, things were not as rosy.  Leaked US diplomatic cables from Robert McCallum, ambassador to Australia, described Rudd as a "control freak" and "a micro-manager."  The week before his Carbon Pollution Reduction Scheme began debate in Parliament, Kevin Rudd made a dramatic attack on climate change sceptics in a speech at the Lowy Institute:

> It is time to be totally blunt about the agenda of the climate change sceptics in all their colours – some more sophisticated than others.  The legion of climate change sceptics are active across the world, and they happily play with our children's future.

> The clock is ticking for the planet, but the climate change sceptics simply do not care. The vested interests at work are simply too great.

> Put more simply: these climate change sceptics around the world would be laughable if they were not so politically powerful – particularly in the ranks of conservative parties.

> The climate change deniers now form the comfortable bedfellows of the global conspiracy theorists – in total bald-faced denial of global scientific, economic and environmental reality.  These arguments – thinly veiled attempts to create a new climate change global conspiracy theory – are now being used in Australia.

> Like the arguments from climate change deniers, these arguments have zero basis in evidence.

Hmmm, zero basis in evidence.

Let's start with what the IPCC forecasts the effects of climate change will be in Australia from AR4:

> Where the analysis has been done for Australia (e.g., Whetton et al., 2002), the effect on changes in extreme temperature due to simulated changes in variability is small relative to the effect of the change in the mean. Therefore, most regional assessments of changes in extreme temperatures have been based on adding a projected mean temperature change to each day of a station-observed data set. Based on the CSIRO (2001) projected mean temperature change scenarios, the average number of days over 35°C each summer in Melbourne would increase from 8 at present to 9 to 12 by 2030 and 10 to 20 by 2070 (CSIRO, 2001). In Perth, such hot days would rise from 15 at present to 16 to 22 by 2030 and 18 to 39 by 2070 (CSIRO, 2001). On the other hand, cold days become much less frequent. For ex-

ample, Canberra's current 44 winter days of minimum temperature below 0°C is projected to be 30 to 42 by 2030 and 6 to 38 by 2070 (CSIRO, 2001).

By 2020, significant loss of biodiversity is projected to occur in some ecologically rich sites, including the Great Barrier Reef and Queensland Wet Tropics.

By 2030, water security problems are projected to intensify in southern and eastern Australia and, in New Zealand, in Northland and some eastern regions.

By 2030, production from agriculture and forestry is projected to decline over much of southern and eastern Australia, and over parts of eastern New Zealand, due to increased drought and fire. However, in New Zealand, initial benefits are projected in some other regions.

By 2050, ongoing coastal development and population growth in some areas of Australia and New Zealand are projected to exacerbate risks from sea level rise and increases in the severity and frequency of storms and coastal flooding.

The first part of this is astonishing! Please take a moment to re-read it. Initial modelling suggested there was no change to the Australian extreme climate, so a second model was used by increasing the temperature of each day. Increasing each day obviously increases the number of extreme days.

These extreme hot days result in the forecasts of climate doom in Australia. Increased drought, increased fires with decreased water and decreased bio-diversity.

And how was this reported in the media? Brisbane's Courier-Mail newspaper headlined in May 2009 that: "Climate change to kill coastal tourist attractions. Super cyclones. Heatwaves. Catastrophic coastal flooding in north Queensland."

The Meteorology Act of 1906 allowed the newly formed Commonwealth Government to centralise the former Colony's function of weather monitoring and forecasting. The Australian Bureau of Meteorology is the only organisation in Australia allowed to forecast the weather. The Council of Science and Industrial Research (CSIRO) was formed by the Government in 1926. It heralds its scientific breakthroughs to include advanced radio astronomy, atomic absorption spectroscopy, biological control of rabbits and plastic banknotes.

These two organisations represent the keystone for climate change data and interpretation in Australia. They jointly release the *State of the Climate* snapshot.[75] It states that all of Australia has experienced warming over the past 50 years. Bureau of Meteorology Director Dr Greg Ayers said the observed changes showed climate change was real. "Australia holds one of the best national climate records in the world. The Bureau's been responsible for keeping that record for more than a hundred years and it's there for anyone and everyone to see, use and analyse."

So why hasn't the climate in Albany, Western Australia warmed over the past 60 years? I don't mean to be too critical, but if the CSIRO and the Bureau of Meteorology make the claim that all of Australia has been experiencing warming, and Albany is in Australia, and we are free to use and analyse the data, why has Albany not warmed?

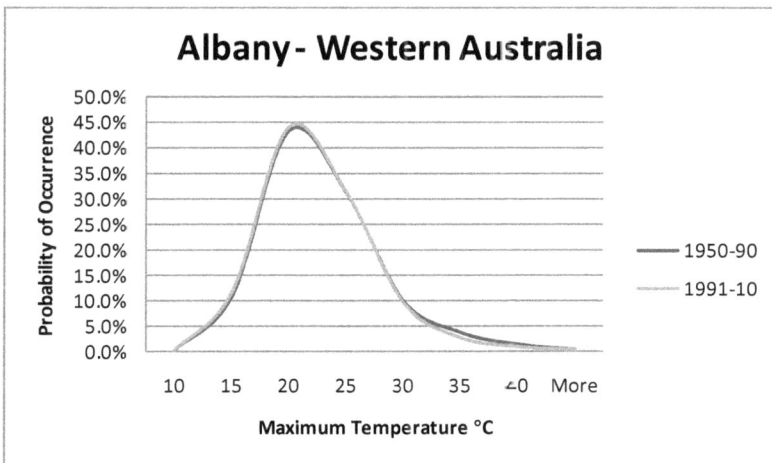

**Figure 88 - The maximum temperature in Albany, Western Australia has not changed.**

The *State of the Climate* footnote says that the "CSIRO and Bureau of Meteorology use scientific modelling based on the laws of physics and thoroughly tested against recorded observations."

Now, you wouldn't want to take on the laws of physics, so I guess they didn't test their claim of warming thoroughly.

What about the measurements for Hobart? Sometimes Australian's forget about the island of Tasmania. Many Tasmanians still believe that their state was deliberately omitted from the map in the opening ceremony of 1982 Commonwealth Games in Brisbane. The organising committee denied this but we have never seen a map without Bass Strait since.

The distribution for Hobart from 1944, Figure 89, shows a climate virtually unchanged. If you examine the same data starting from 1960, the common reference date for anomaly data used by the CSIRO and the IPCC, the fit between these distributions is even more exact. Dr Greg Ayers is correct is his assertion that Australia holds one of the best national climate records in the world.

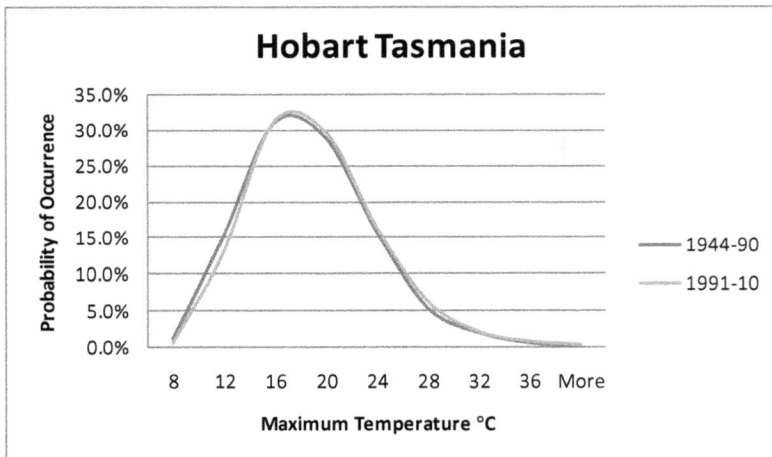

**Figure 89 - Climate in Hobart Tasmania has been virtually unchanged since 1944.**

Temperatures for Sydney start from 1858 and Adelaide from 1887, but it is Melbourne that wins with records from 1855. This is very impressive since the city was only founded in 1835.

These records show the climate has changed in Sydney.

Founded in 1788, Sydney had approximately 40,000 people in 1859 when the measurements began. The warmest day recorded was on 14 January 1939. It was 44.8°C. The coldest temperature recorded was also in the 1930s, 2.1°C on 22 June 1932. Yet these are just facts for a trivia night. Individual temperatures do not mean much when it comes to analysing a climate. The climate of Sydney is very temperate. It doesn't snow, direct cyclone hits avoid the city. It is a lovely place to live. Just ask any of its 4,504,469 residents today.

The fastest growing suburb – Canada Bay.

One hundred times growth in population has produced climate change. The shift in the distributions, comparing 1859-1900 and 1991-2010, as show in Figure 90, is interesting for two reasons. Firstly, 4.5 million people have thrived in a city 'ravaged' by the projected impacts of climate change. The Bureau of Meteorology calls this urbanisation effect. And secondly, and significantly, there has been no change in the profile of hot days over 33°C.

Also, when you examine the climate from 1960-90 and compare this to the climate from 1991-10, the effects of carbon based climate change are nothing compared to adding 100 times more people.

Of interest, the profile of minimum temperatures in Sydney has only had minor variation since 1960. Refer to Figure 92.

Move just 70 kilometres to the northwest and you find the city of Richmond. Once a country town, the urban expansion of Sydney is slowly gobbling it into a suburb. Richmond's temperature distribution has only a few days different at 12°C and the remaining climate has not changed one fraction since 1940. See Figure 93. This is fairly solid evidence that aside from the urbanisation factors, the climate in the Sydney region has actually not changed.

Now Richmond has a somewhat similar climate to Appin. We have come full circle.

We now get on the Hume Highway and travel the 963 kilometres southwest to Melbourne.

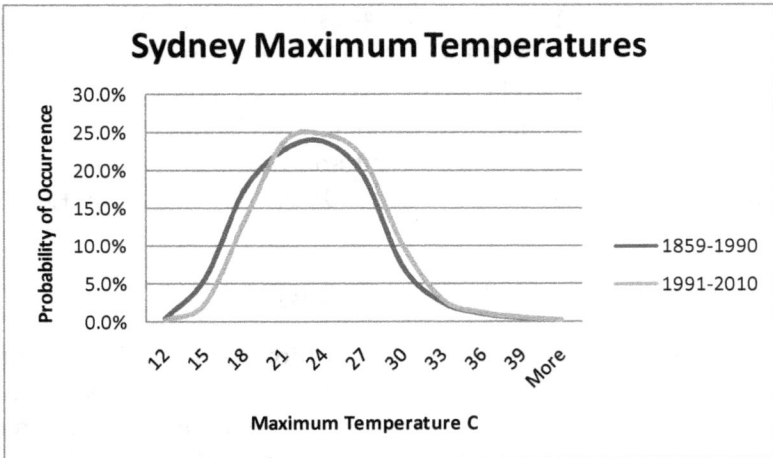

**Figure 90 - One hundred times growth in population and the city of Sydney shows a rather large shift in the climate. Comparing the maximum temperatures from 1859 to 1900 against 1991 to 2010.**

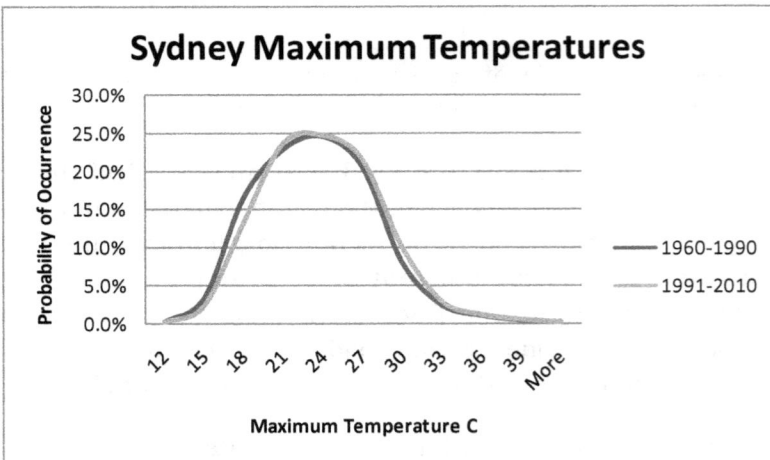

**Figure 91 - The distribution difference between 1960-90 and 1991-10 yields a very similar climate in Sydney.**

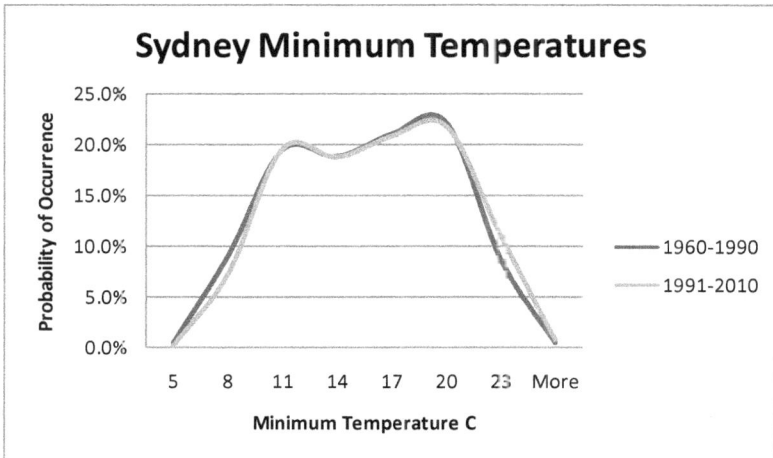

**Figure 92 - The distribution of minimum temperatures for Sydney has also not changed significantly since 1960.**

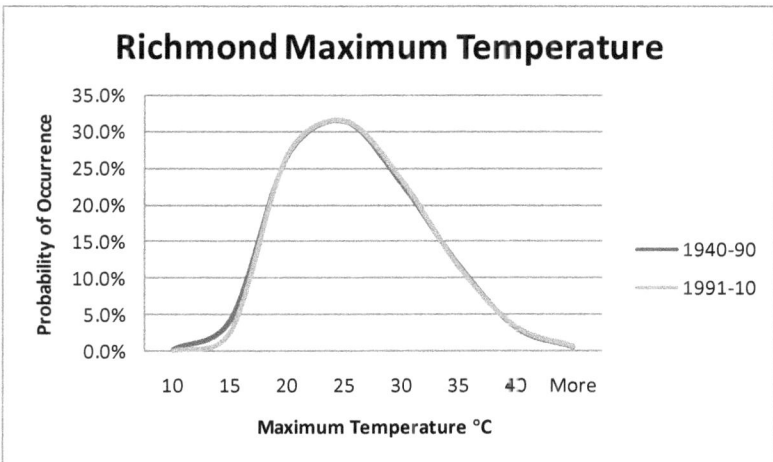

**Figure 93 - Richmond NSW, about 70 kilometres northwest of Sydney, has had virtually no climate change since 1940.**

Melbourne is famous for having four seasons in one day. It is one of those places where a single temperature doesn't do the climate justice. It actually receives about half the rainfall of Sydney, yet is has the reputation amongst Sydneysiders as being a cold, rainy and dreary place. Melbournians however don't share this view and really love their climate.

From the time that records began in 1856 until 1990, Melbourne's maximum temperatures have remained virtually unchanged. This was very surprising. The distribution is shown in Figure 94.

Yet since 1991, the maximum temperatures have changed significantly. Melbourne now has less cool days and more warm days. Refer to Figure 95. Unlike Sydney, Melbourne has experienced some climate change.

Now is having less cool days and more warm days a bad thing? Has global warming turned Melbourne into a nicer place for its 4 million residents to live?

But what is most significant in the data is that the changes to the extreme hot days have not occurred.

I still can't believe how simple a model the CSIRO used to determine temperature extremes. They added a projected mean to each day and made the claim that Australia will experience more extreme temperatures. I know that the CSIRO is Australia's preeminent science body. They invented the plastic bank note. But in this case, they got it totally wrong.

This is a huge mistake that has made it past all the scientific hurdles and into the IPCC AR4 assessment report.

Melbourne now sees 27.3 less days a year when the maximum temperature was 15°C and has replaced these days with warmer days between 20-30°C not hot days over 35°C.

The incorrect assertion that extreme temperatures are increasing have been allowed to influence the reporting in the media for a decade. The increase in extreme hot days is the basis for increased bush fire, drought and damages to our ecosystem. This was a huge mistake.

Yet the changes to Melbourne's climate have not been enough to change it into Sydney. Melbourne wins with 55,479 temperature measurements since 1859, 29 more days measured than Sydney. But Sydney does have a better climate, with substantially less cold days, more warm days but less hot days, as shown in Figure 97.

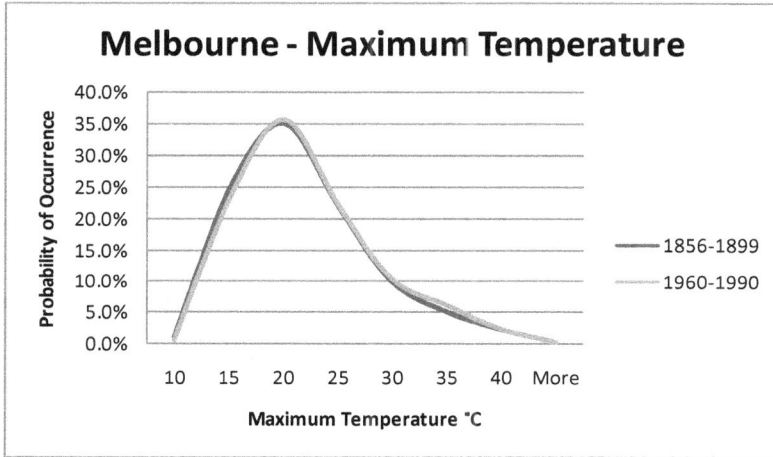

**Figure 94 - Melbourne's maximum temperatures have been virtually unchanged since 1856.**

There isn't much of a difference between the ecosystems in Melbourne and Sydney. Comparing the difference in climates between Melbourne and Sydney wasn't just done for fun. Since Australia's biodiversity is forecast to reduce dramatically from 2020, this comparison shows the resilience of the ecosystem in the temperate regions.

Moving to the west is the city of Adelaide, its temperature records start from 1887. Adelaide has not experienced the same massive population growth like Sydney and Melbourne; it only has 1.2 million people. Its temperature records show a slight difference in cold days with a similar increase in warm days at 20°C. Again, there is no increase in extreme heat. This is comparing 123 years of data.

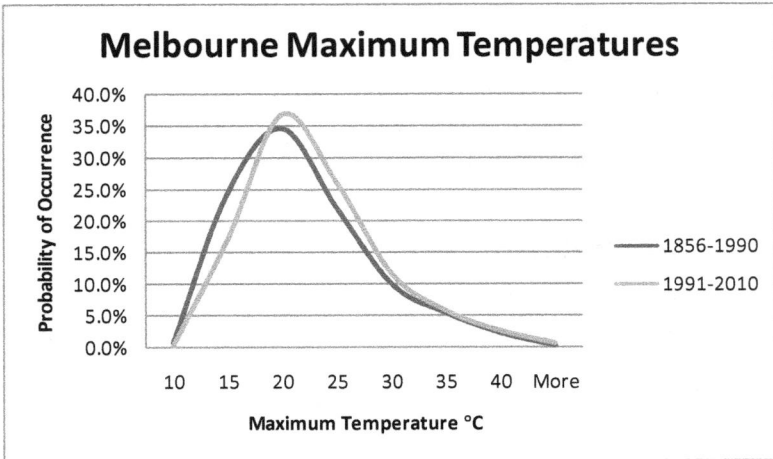

**Figure 95 - Since 1991, the maximum temperatures in Melbourne has changed significantly. There are far less cool days and more warm days. The one thing that hasn't changed is there are no more extreme days.**

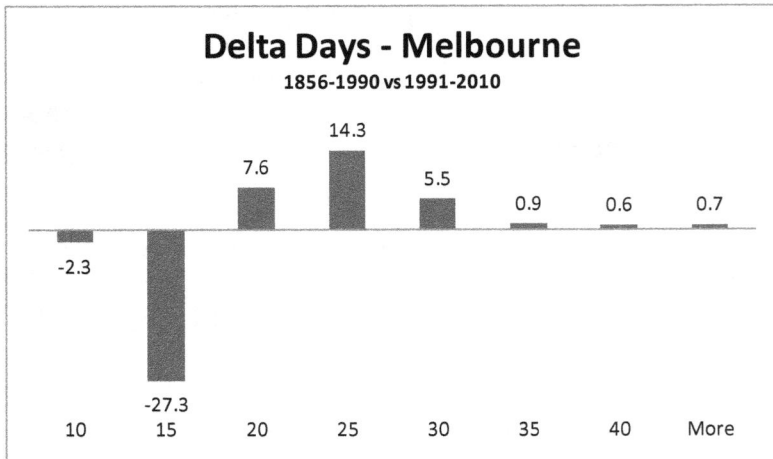

**Figure 96 - The climate has changed in Melbourne with less cold days and more warm days. However, the number of extreme days over 35 has stable.**

## Melbourne vs Sydney
### Maximum Temperatures - 1859-2010

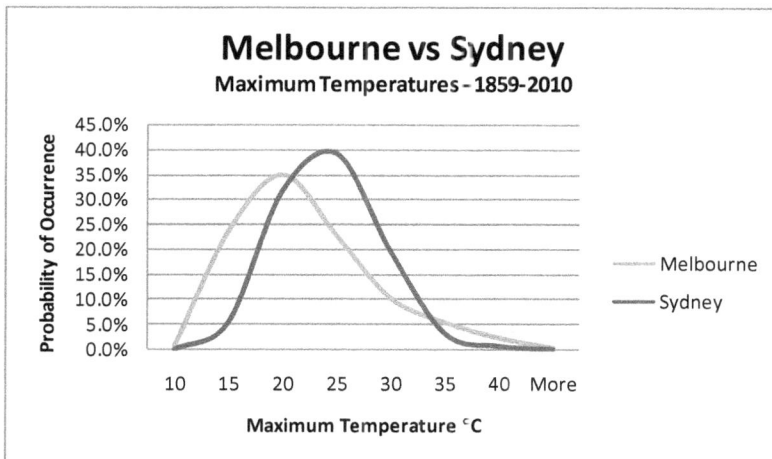

**Figure 97** - The climates of Sydney and Melbourne are very different yet the ecosystems between the two locations are virtually identical. This fact casts some doubt on the CSIRO forecast that Australia's ecosystems are doomed by 2020.

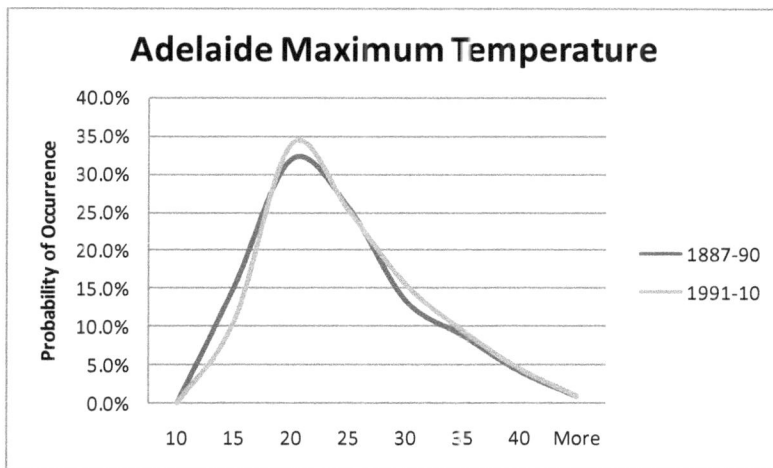

## Adelaide Maximum Temperature

**Figure 98** - Adelaide has experienced very little climate change and no increases to the extreme days incorrectly forecasted by the CSIRO.

141

At this point, we take a trip across the Tasman Sea to New Zealand; the IPCC group Australia and New Zealand into the same climate zone. The Kiwi's actually forecast some benefit from climate change.

The town of Invercargill is famous for Burt Munro and his highly modified 1920 Indian motorcycle setting the land speed record for a 1000cc bike. The record, set in 1968 while Burt was 68, is still held today. The climate of Invercargill has actually cooled slightly since 1949, maybe because Burt isn't around anymore. Refer to Figure 99.

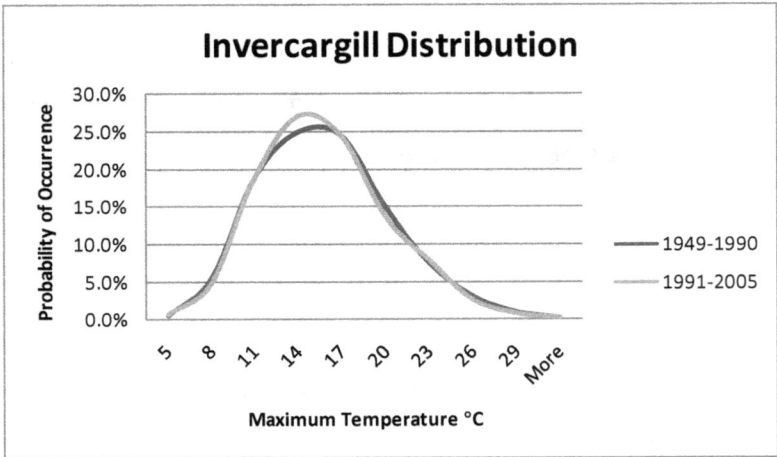

**Figure 99 - The temperature at Invercargill New Zealand has cooled slightly, with more days at 14°C and less in the 20s.**

It seems that many of the temperate climates of Australia and New Zealand are not yet experiencing the effects of climate change.

Yet the impact on biodiversity has been forecasted to hit worst in the tropics, so let's take a trip up north. Australia's Great Barrier Reef consists of about 3000 individual reef systems in the Coral Sea, extending thousands of kilometres from Rockhampton in the south to past Cairns in the north.

Ever since I arrived in Australia in 1985, I've heard doomsday projections regarding the Great Barrier Reef. First it was pesticides and nutrients from agriculture, than it was salinity. We had copper, nitrogen and phosphorus contamination followed by increased levels of mud and sediment. Flooding and cyclone activity were forecast to cause unstainable disturbances, taking decades for the coral to recover. Oil spills from coastal shipping and the discovery of oil rich strata and the possibility of exploration, lead to a major concern over oil pollution. Shipping also could introduce contaminated waste from ballast water, toxic paint from ship hulls could leach into the water as well as the possibility of ships running aground and destroying the coral. Over fishing and illegal poaching can take its toll on adequate life and the balance of the reef's ecosystem. The crown-of-thorns starfish could consume the reef and its larvae thrive from human pollution. And did I forget to mention ocean acidification and infectious plagues and disease in the coral?

It's a wonder the reef survived the past 6000 years.

The concern now with reefs is from coral bleaching.

The Great Barrier Reef Marine Part Authority explains that coral bleaching occurs when the coral expels a marine algae called zooxanthellae. This algae produces excess nutrition which the coral consumes. Without it, the coral, a living animal that just happens to look like a plant, can die. Studies suggest that high water temperatures of only 1.5-2.0°C, lasting for six to eight weeks, is enough to trigger bleaching. When conditions return to normal, the coral will recover, but persistent high temperatures of more than eight weeks can cause the coral to die.[76]

Dr Ray Berkelmans lead a team from the Australian Institute of Marine Science to examine the coral bleaching events in 1998 and 2002. Their report proved interesting. They found that the "maximum surface sea temperature over any 3-day period during the bleaching season predicted bleaching better than anomaly-based surface sea temperature and that short averaging periods predicted bleaching better than longer averaging periods."[77]

Finally, someone in the climate change game using raw data rather than long term anomaly averages.

The report claimed that 2002 was the worst bleaching event on record for the Great Barrier Reef. Mass bleaching events do not necessarily affect all coral in the same region. In 1998 and 2002, 18% of the coral studied was bleached strongly.

An unexpected result of the study found that short periods of high temperature proved highly stressful to corals rather than long term high temperature water. "Latitudinal variation was suggestive of long term adaptation in coral communities" - in other words, – corals can live in hotter water, just not sudden increases to water temperatures. This said, the model based solely on sea surface temperatures correctly predicted bleaching in only 73.2% of coral studied. It also found that inshore reefs were more vulnerable to bleach than offshore reefs, even at the same water temperature.

Their model was then adapted to handle the IPCC temperature forecasts from global warming and concluded that a 3°C increase in sea surface temperature would lead to 100% coral bleaching - "In the absence of acclimatization/adaption, [reefs] are likely to suffer large declines under mid-range IPCC predictions by 2050."

The first reported coral bleaching was in 1911 at Bird Key reef in the Florida Keys. In 1929, a bleaching event was noted in the now popular Low Isles of the Great Barrier Reef near the tourist destination of Port Douglass. Further bleaching events were not recorded until 1961, when it was again noticed in the Florida Keys. Since 1979, reported bleaching events have increased and this trend has been linked to climate change.[78]

I went diving on the Great Barrier Reef in 2004 and personally witnessed coral bleaching. It is very unattractive; white coral devoid of any colour. It is not clear if bleaching occurred in the past and was not reported. It is interesting to note the increase in reported events of bleaching coincide with the popularity of snorkelling and scuba diving. Yet you would have to assume that bleached white coral would have been noticed in the past.

The CSIRO scientific journal, Marine & Freshwater Research, published Dr Ove Hoegh-Guldberg's paper, *Climate change, coral bleaching and the future of the world's coral reefs,* in 1999. This paper seems to be the key reference in today's claims that global warming threatens the world's coral reefs.

While the paper found strong bleaching episodes coincide with periods of high surface sea temperatures and changes to the El Nino periods, temperature alone could not explain mass bleaching events. It found that there was "often a graduation of bleaching intensity within colonies, with the upper sides of colonies tending to bleach first and with the greatest intensity." Given that temperature is unlikely to differ between the top and sides of a coral colony and bleaching does not correlate perfectly with temperature, other factors should be considered. Bleaching can also differ between colonies that are located side by side.

One theory is that unusually calm seas allow elevated levels of the sun's ultraviolet radiation to penetrate the coral. This fits the model as to why the upper sides of colonies tend to bleach first and that bleaching occurs more often on inshore reefs; the shallower water allows more of the sun's rays to penetrate.

In 1993, experiments on reef corals with artificial manipulation to change the levels of ultraviolet radiation did trigger a bleaching response in the test coral.[79]

However, since there is a complete absence of reported bleaching in cooler waters, some scientists believe that warmer temperatures and increased ultraviolet radiation need to be combined to trigger bleaching.

Yet most of the references, including the Great Barrier Reef Marine Park Authority, say "the primary cause of coral bleaching is high water temperature."[80]

Globally, 1998 was a terrible year for coral bleaching – it is estimated that 16% of the world's reefs died. The Seychelles bleached 90% of its coral.

Yet the climate in the Seychelles has experienced only minimal changes. Because of dataset limitations, the years from 1972-89 were compared to 1990-2003.

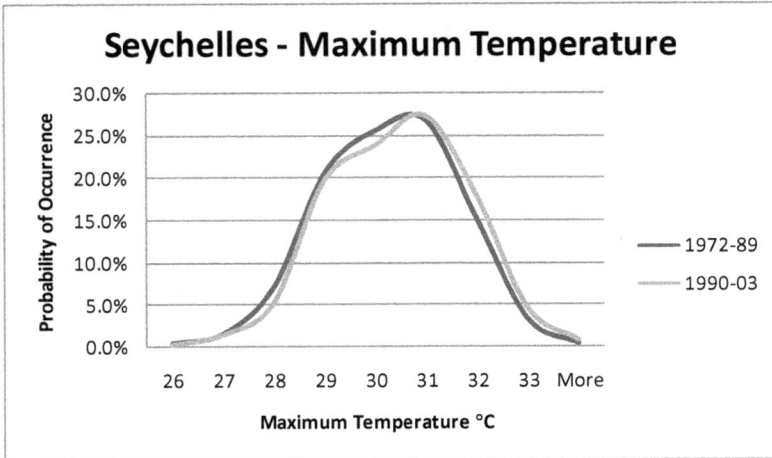

**Figure 100 - The maximum temperature in the Seychelles has remained very similar.**

The Seychelles experienced a very warm season in 1998. (Refer Figure 101). It was at the extreme of the measurements in the 1970s. Up to 22% of shallow water coral in less than 10 meters of water died but deeper water deaths were negligible, despite similar water temperatures.

In Australia, temperatures in the Great Barrier Reef coastal cities of Cairns, Townsville, Mackay and Rockhampton, have all experienced significant climate variation.

While all these sites have changed, the city of Rockhampton saw the least amount of variation based on records since 1940. See Figure 106.

Cairns and Townsville have changed since 1943. There are less cold days and more warm days, but no more extreme hot days. The maximum temperature distribution is shown in Figure 103 and Figure 104.

On the other side of the Coral Sea, similar changes to the maximum temperature distribution were seen for New Caledonia. See Figure 107.

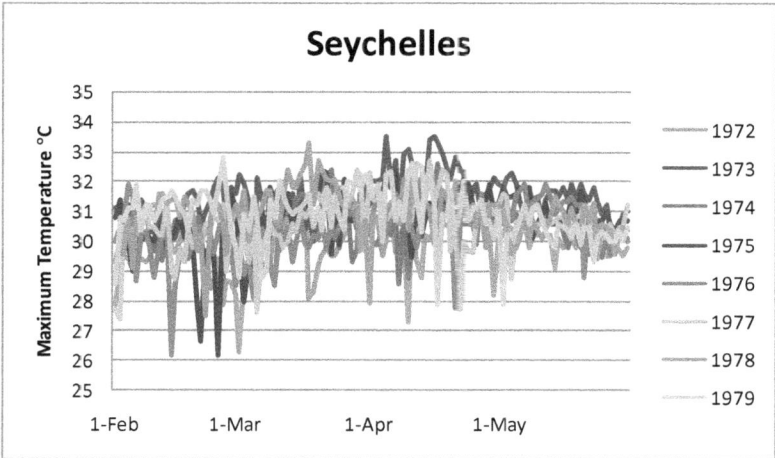

**Figure 101 - The maximum temperatures for the Seychelles warm season, from February to May, show temperatures in the range of 30-32°C are most common.**

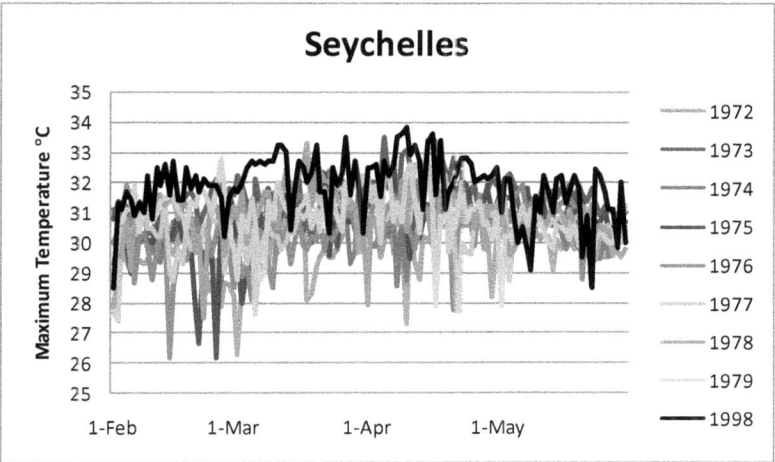

**Figure 102 - During the coral bleaching of 1998, the temperatures were unusually high, but still within the range of previous non-bleaching years.**

These coastal cities have the most temperature variation noticed in the locations analysed. The strange thing - inland from the coast, the same temperature variations are not evident. For example, the outback city of Longreach, home to the *Stockman's Hall of Fame*, and the city of Charters Towers, saw only a slight change in the temperature profile. Refer Figure 108 and Figure 109.

Did this change in climate cause the coral bleaching along the Great Barrier Reef in the summer of 1998?

To examine this, the daily summer temperature for the 1950's is analysed. A decade when coral bleaching was not witnessed. See Figure 110.

On 10 February 1998, reports were received that the surface sea temperatures had warmed considerably over the past few weeks. Four days later, the first reports of bleaching were received. By 27 February, heavy inshore reef bleaching occurred and by mid-March, the entire length of the Great Barrier Reef had some form of bleaching. [81]

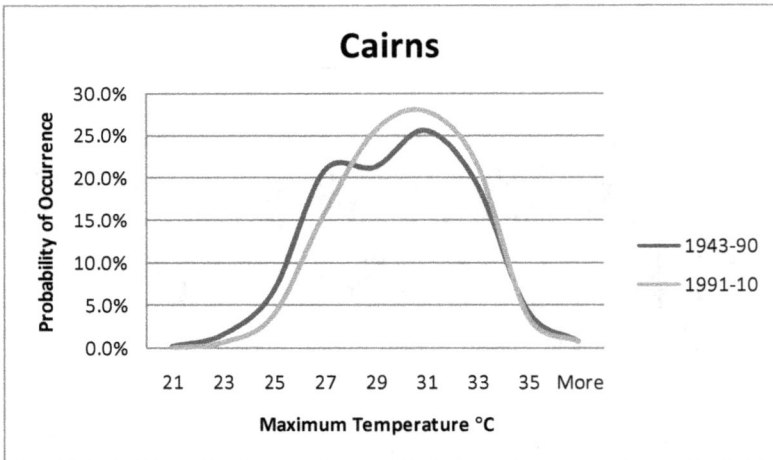

**Figure 103 - The climate of Cairns, in tropical north Queensland has changed since 1943. More cool days have been replaced by more warm days. Unlike the other Great Barrier Reef locations, there has been no change to the extreme hot days in Cairns.**

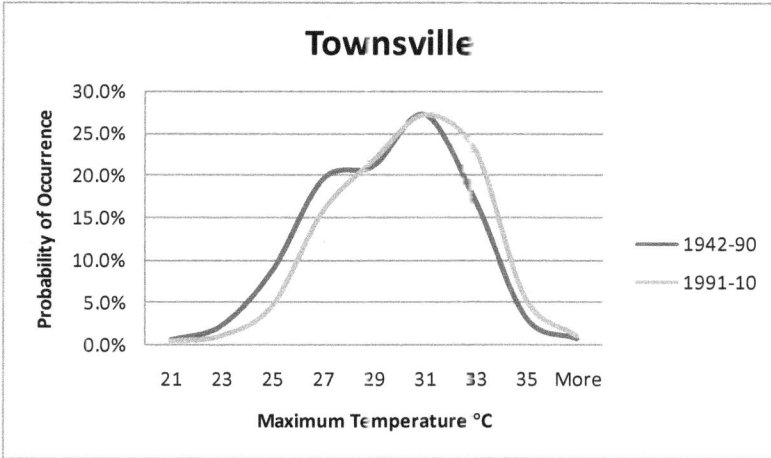

**Townsville**

Maximum Temperature °C

Probability of Occurrence

1942-90
1991-10

**Figure 104** - Like Cairns, Townsville has experienced significant changes to its climate since 1942. Less cool days have been replaced by more warm and hot days. Townsville has experienced more extreme hot days.

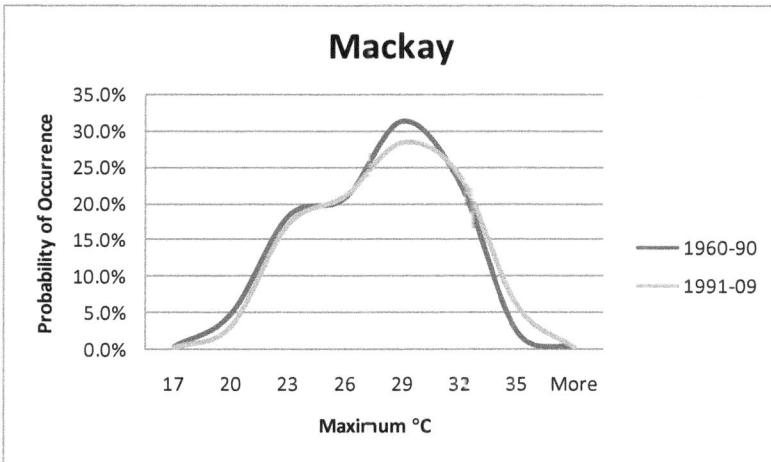

**Mackay**

Maximum °C

Probability of Occurrence

1960-90
1991-09

**Figure 105** - Mackay has experienced less hot days and replaced these with more extreme hot days. The cool day profile is very similar.

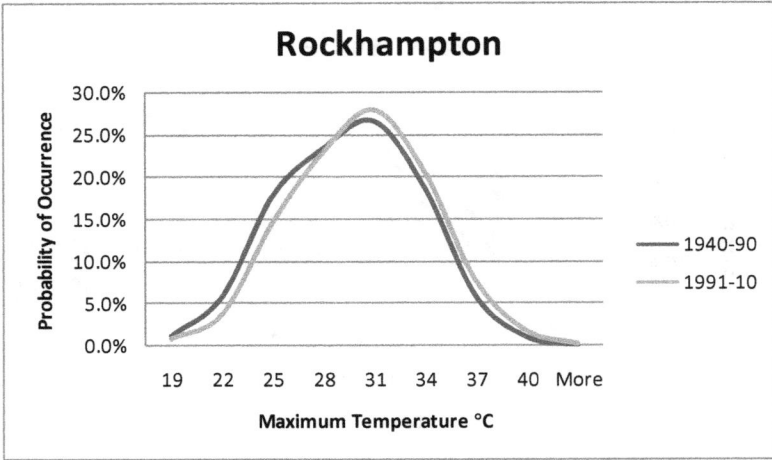

**Figure 106 - The city of Rockhampton has seen the most consistent climate in the Great Barrier Reef coastal cities, with some changes to the cool days leading to an increase in warm days and a slight variation in hot days.**

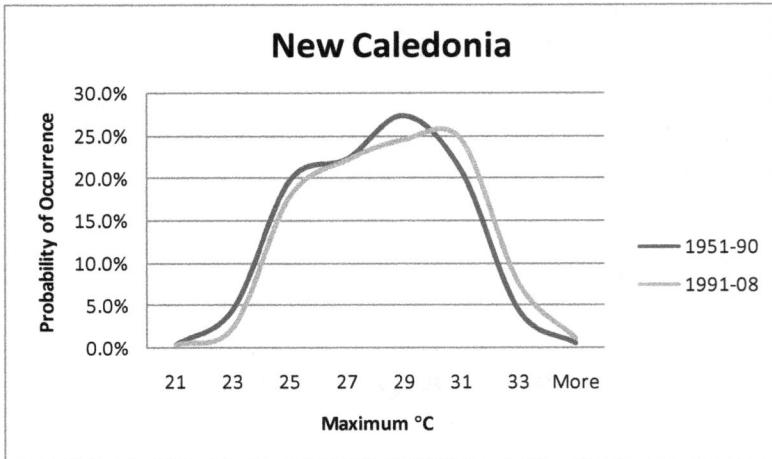

**Figure 107 - On the other side of the Coral Sea, New Caledonia has seen a considerable change in its temperature distribution.**

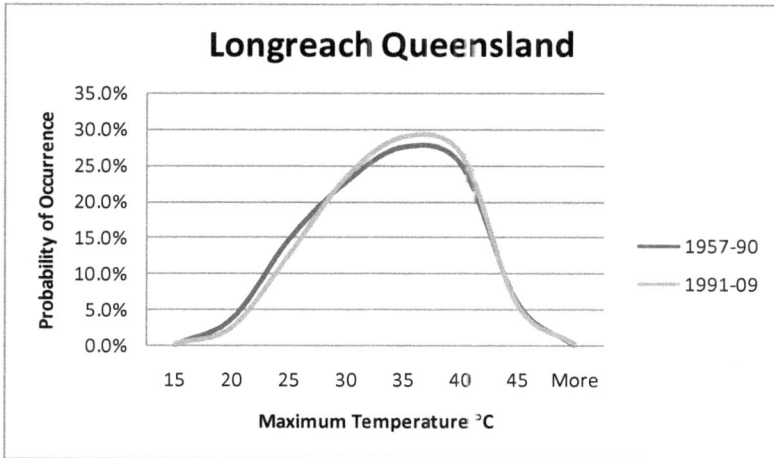

Figure 108 - In Longreach, the temperature profile for this outback town saw very little change in the distribution of maximum temperatures. The same thing is true for Charters Towers, Queensland.

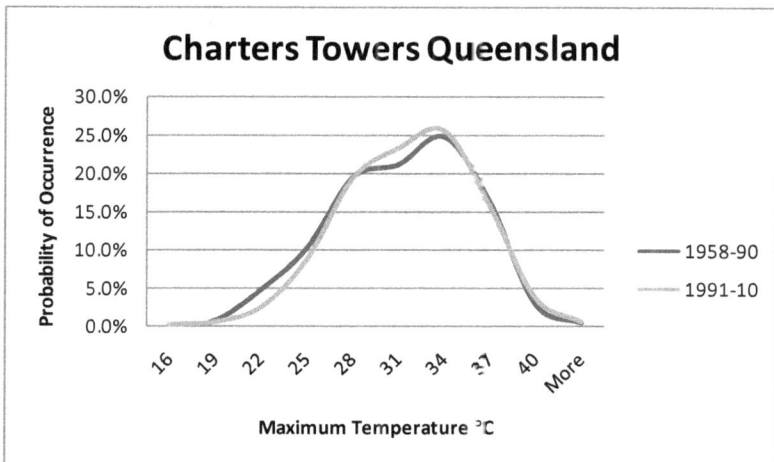

Figure 109 - Like Longreach, Charters Towers has not show obvious signs of warming.

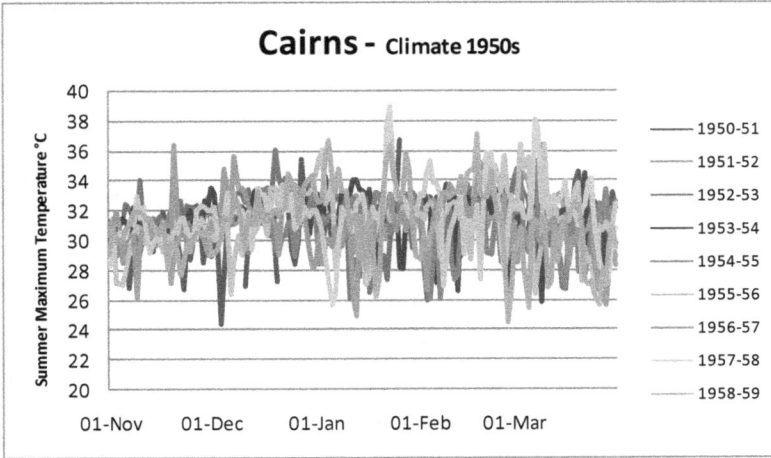

**Figure 110** - **The summer temperatures for Cairns in the 1950s shows some extreme variation, with temperatures over 38°C.**

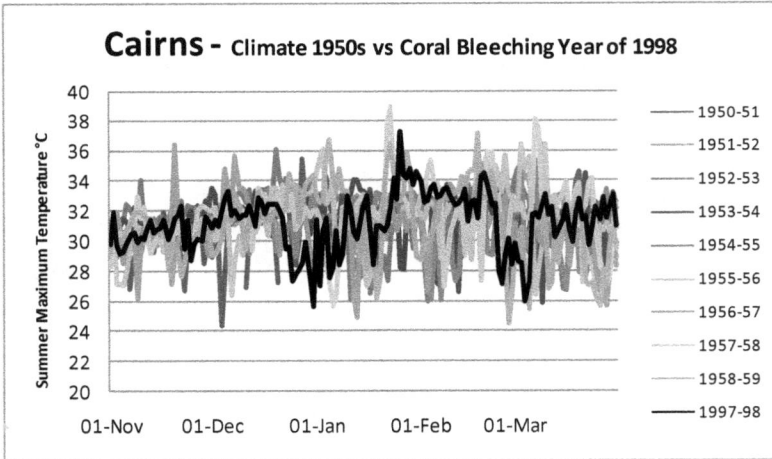

**Figure 111** - **The summer of 1997-98 in Cairns saw no increase in air temperature that could be attributable to the major coral bleaching event of 1998.**

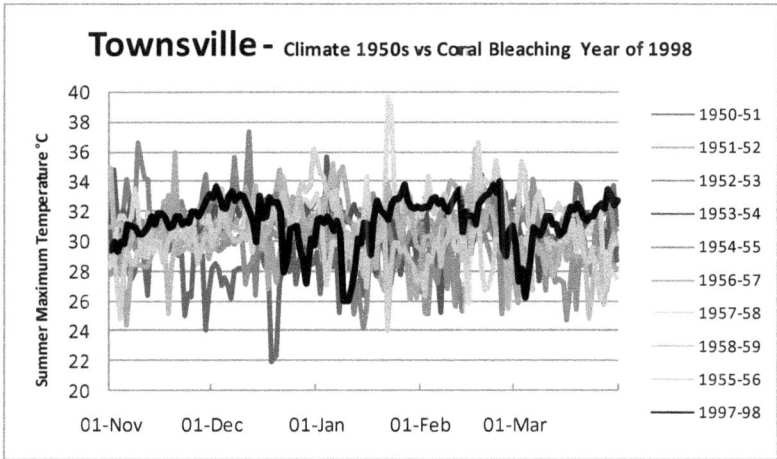

**Figure 112 - In Townsville, the summer of 1997-98 was very typical of a summer in the 1950s.**

The low temperature in Cairns of 25.6°C on 31 December 1997 saw a colder January than in the 1950s. A sharp rise to 37.2°C on 26 January was still lower than individual days in 1957-58. Yet one day at 37°C is not going to raise the Ocean temperature; especially when the rest of the summer was so consistent and the lead up days were cooler than normal. (Refer Figure 111). Similar data for Townsville (Figure 112) saw a very similar climate to the 1950s.

It is difficult to see from this data that temperature alone caused the bleaching of 1998.

What is very interesting is that in the IPCC climate change model, increased temperatures are supposed to bring additional rainfall. This has not occurred.

Just to the north of Cairns, in the suburb of South Mossman, rainfall measurements have been taken since 1911. For some unknown historical reason, rainfall and temperature measurements are not taken at the same location. We saw this at the Cataract Dam in Appin and the same thing is true for North Queensland. But we are close enough for the analysis.

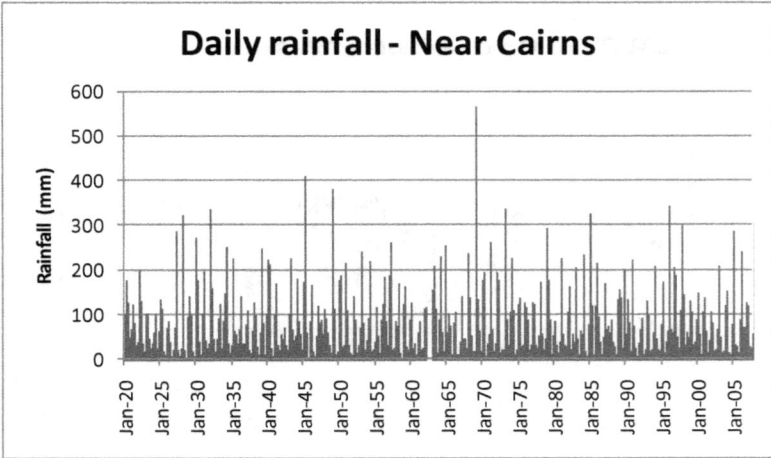

**Figure 113** - The daily rainfall from 1920 shows a very consistent rainfall profile. Data exists from 1911, but technology limitations enable only 32,000 to be plotted.

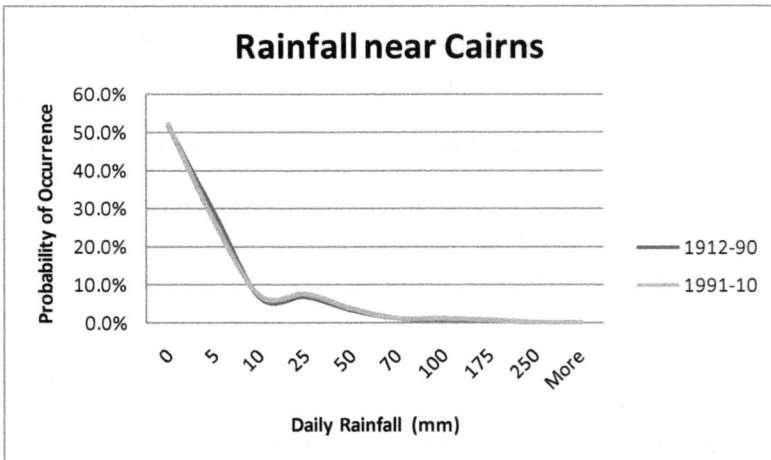

**Figure 114** - Rainfall from 1912 to present has remained unchanged near Cairns. The Global Warming model forecasts additional rainfall and an increase in rainfall intensity. Neither of these has been recorded near Cairns.

## Rainfall near Townsville

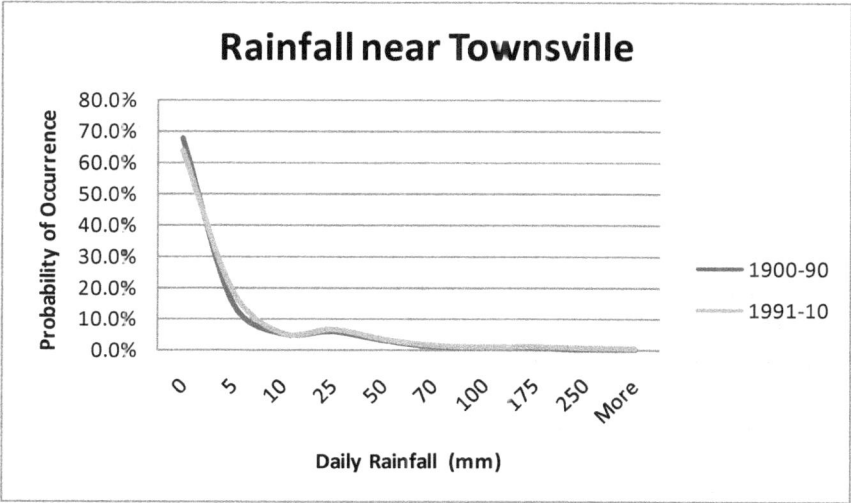

Figure 115 - Rainfall distribution near Townsville shows no significant increase or change in rainfall patterns for the past 110 years.

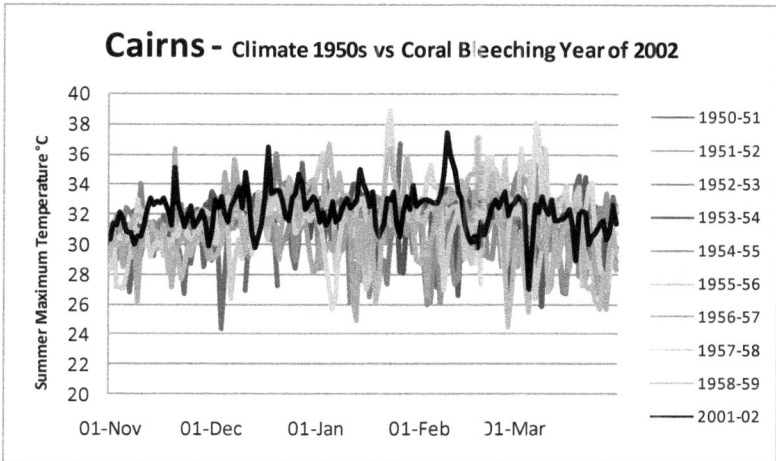

Figure 116 - The summer of 2002 in Cairns was warmer than 1998. But while the maximum temperature was at the higher end of those recorded in the 1950s, it was still within the general ball park.

The rainfall near Cairns has not varied at all since measurements were taken. See Figure 114. The same is true for rainfall at the Macknade Sugar Mill near Townsville. See Figure 115. These measurements start in 1900. For rainfall not to increase with higher temperatures puts these results in direct conflict with the IPCC climate change model.

Interestingly, in 2002 the Great Barrier Reef experienced another major bleaching episode. While warm temperatures existed all along the coastal cities, it is again not clear these increases in air temperatures were so warm as to damage the reef.

To me, this data shows that bleaching needs a combination of water temperature and solar radiation; a conclusion suggested by Dr Ove Hoegh-Guldberg in the following extract but consequently ignored when citing his paper:

> Certainly, the observation that corals bleach on the upper surfaces first during exposure to elevated temperature argues that the quality and quantity at solar radiation are important secondary factors.

Let's review the results:

1. Sea temperature alone failed to forecast 26.8% of coral bleaching.
2. The upper sides of colonies tend to bleach first and with the greatest intensity.
3. Deeper water corals in the same water temperature didn't bleach or die to the same extent.
4. Bleaching also differed between colonies that are located side by side.
5. The air temperatures in the Great Barrier Reef and the Seychelles during the bleaching events of 1998 were warm but not extreme.

Similar to the melting of ice in the arctic, air temperature alone has not changed enough to influence the bleaching of the coral. Sea currents likely play a vital role in bleaching.

Changing sea currents could also be the cause of the costal temperature variations which were not seen inland or in other parts of Australia or New Zealand.

For example, the tropical temperatures in Darwin, on the top end of Australia, have changed little since 1960. See Figure 117.

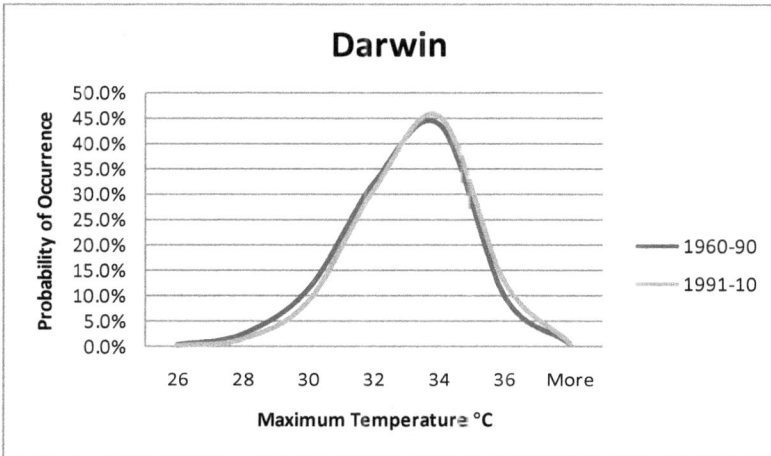

**Figure 117 - The temperature distribution for Darwin does not display the massive changes found in Queensland.**

If the sun's output and changes to the ocean's currents play a role in climate change, then the IPCC's $CC_2$ centric model comes into serious question. The changes observed from the raw data in north Queensland suggest factors other than $CO_2$ levels are involved.

Some good news. The remote Aldabra Atoll in the Seychelles is largely unaffected by human activity – namely no overfishing and a lack of pollution. In the five years following the coral bleaching of 1998, no significant changes in total fish species or diversity were found. This reef system has proved tolerant to bleaching related changes.[82]

Maybe the future of the Great Barrier Reef is simply less exploitation?

Kevin Rudd is from Queensland. His popularity in his home State was vital in securing the 2007 election. But like the Americans, Rudd's own party felt he was a non-consultative micro-manager who had to be stopped. Who cared about voter popularity? So he was dumped as Prime Minister and replaced by his deputy, Julia Gillard; Australia's first female in this leadership role.

In an attempt to consolidate her position, she called an election. She must have forgotten that elections do consider voter popularity. She lost the lead her party gained in 2007 and with only 38% of the primary vote, she won only 72 of the 76 seats needed to form Government.

The winners on the night were the Australian Greens. With only 11.7% of the vote, they secured the balance of power in the Australian Senate from 1 July 2011.

It seems that Julia is very good at back room dealings. She was able to solicit the support of four key independents to form Government with the promise of spending $43 billion on a National broadband network to connect regional Australia.

Her Government also did a deal with the Green party for their support in the senate - the climate change debate in Australia is far from over.

As for Kevin Rudd? While public opinion polls still show that he is Australia's preferred prime minister, the popular "control freak" was given back to the Americans so to speak. He was made Australia's Foreign Minister.

# Chapter 13
# HYPER-INFLATED MARKETING

Marketing can be a very powerful tool.

When I was a kid, about 12, I remember defending the Vietnam War to a group of class mates. "But they attacked our ships," I told them.

I was referring to the Gulf of Tonkin incident. The American people and the western world were told of two incidents when a group of North Vietnamese in gunboats attacked the USS Maddox while on a friendly patrol mission.

What we didn't know was that our ship was the aggressor. It was conducting covert operations for the CIA. In 2005, declassified documents showed this and that the second incident in particular, never occurred.

At the time, the outcry from these incidents allowed President Lyndon B. Johnson to pass the *Gulf of Tonkin Resolution*, which granted the President authority to support any Southeast Asian nation that was threatened by a communist aggressor, without a declaration of war. Wasn't it the lack of war declaration that upset us most about the Japanese attack on Pearl Harbour?

But there was a red under every bed and the *first causality of war is truth.*

This expression was attributed to US Senator Hiram Johnson in 1918. His actual quote was "the first casualty when war comes is truth." Yet the saying goes back much further than that, when the Greek poet Aeschylus wrote in 500BC, "in war, truth is the first casualty."

The use of *war truth* is actually part of a complex and highly specialised form of marketing that Governments have honed to perfection over the centuries.

There is much more to marketing then selling soda-pop or chocolate bars. Marketing is the skill which allows a targeted message to reach the masses and, more importantly, generates them to action. A powerful extension of this is *Hyper-Inflated Marketing.*

*"Are you now or have you ever been a member of the Communist Party?"*

Thousands of Americans had their lives destroyed when they were berated with this question during congressional hearings, accused of being a communist. Communism is now almost a distant memory, represented today by headlines of North Korean military aggression or the Maoist insurgency destabilising Nepal. It's easy to forget our fears of Chinese and Soviet aggression; yet concerns about these systems taking over the world reached fever pitch in the 1950s and continued well into the 1990s. We now look back and almost chuckle at *McCarthyism*, yet millions died in the battlefields of Korea and Vietnam to halt the Red Menace. Continuing into the 1980s, we funded and armed the radical Islamic movement fighting the communist Soviets in Afghanistan only to have their anger extended towards us after the Russians withdrew. In hindsight, was the Soviet invasion of Afghanistan an attempt to curb radicalism rather than expand communism? At the time, we only saw the communist threat; these shades of grey were deliberately not promoted. Now the Russians, Americans and Chinese are jointly facing the threat of Islamic extremists in what has become a very complex struggle.

War, nationalism and marketing are not generally combined in common thought. Yet *winning the hearts and minds* has become the catch phase in modern war, as it is in marketing. Building values, establishing effective communication strategies, promoting continued loyalty and ensuring action. These are the core principles of marketing and modern warfare.

The skill that Governments have obtained is an ability to take the standard marketing model to a new level; and more recently, these *hyper-inflated marketing* principles have migrated out of the exclusive sphere of Government control.

So let's start by explaining the principles.

### *Principles of Hyper-Inflated Marketing:*

1. Be convinced of the common good.
   - *Right* outweighs *Truth*

2. Consequences of inaction are tragic.
   Benefits can be wonderful.

3. Promote risks to the extreme.
   - Simple messages
   - Targeted experts

4. Eliminate critical analysis
   - Avoid scrutiny

## Be convinced of the common good

The starting point here is vital and cannot be undervalued. There has to be a passionate commitment that what needs to be accomplished is for the benefit of the common good. There might be a few blemishes in the fruit, but consuming it is good for all. This concept dates back to Aristotle.

So how did a nation, whose core beliefs are so enshrined that they are even recited by school children every morning in a pledge of allegiance - *"Liberty and Justice for all"* - intern every person who looked Japanese?

At the time, winning the war against Japan was more important than protecting the civil liberties of 110,000 people. In 1944, the US Supreme Court even upheld the internment and it was not until 2007, after decades of denial, that evidence came to light that the US Census Bureau provided the US Secret Service with the names and addresses of the Japanese-Americans.[83]

The Americans were not alone. Australia also interned Italians and Germans in the outback cities of Orange and Hay in New South Wales during the war.

I guess they received better treatment than the 500,000 Volga-Germans, who were interned to Siberia by the Russian Government at the same time.

Do you think times have changed? Sixty years later, the same strategy is still being utilised. Detainees are sent to the US military base in Guantanamo Bay Cuba, considered outside the reach of the U.S. legal jurisdiction, protecting its citizens from the threat of terrorism.

Very often the common good decisions are valid. Would the world be a better place today under Nazi rule? Of course not! But in the shades of grey that influence other elements in our lives, who determines the common good can be open to debate.

## Consequences of inaction are tragic

Let's go back to our Vietnam War example. We were told that inaction to stop communist aggression would lead to a world dominated by the Chinese. What little did we know?

The most difficult part in effective marketing is to solicit action. How many advertisements have you watched where you have done nothing? Most? In *hyper-inflated marketing*, the cost of your inaction needs to be tragic. This could be as simple as your death or possible capture. Or life under a political system that deprived you of your liberties.

At times, tragic inaction can be replaced or combined with wonderful benefits.

For example, in 2003 President George W Bush told us that a "democratic Iraq would lead to more liberalized, representative governments, where terrorists would find less popular support, and the Muslim world would be friendlier to the United States."[84]

The Iraq War provides us with an excellent example of combining wonderful benefits with tragic consequences. National Security Advisor Condoleezza Rice also told us that "much as a democratic Germany became a linchpin for a new Europe . . . so a transformed Iraq can become a key element of a very different Middle East in which the ideologies of hate will not flourish." While Secretary of

State Colin Powell told the world that Iraq possessed weapons of mass destruction. Wonderful benefits with tragic consequences.

## Promote risks to the extreme

Risks exist in anything we do. If you focused on risk, you might never get out of bed. But developing bed sores could be hazardous to your health as well, so you had better get up.

It is easy to promote risks to the extreme. A very good example of this is "*Stranger Danger.*" Researchers from the University of Western Australia examined parental anxieties and found fears about stranger danger did not match up with crime figures. Associate Professor Lisa Wood said, "we found the impact of parental fear on cotton-wool kids is leading to less active lifestyles and increasing obesity levels."[85] It seems that in the rare event the children are abused, it is most often by a friend or family member. So strangers are actually less risky than family. So instead of accepting the risk of letting our children walk to school, we let them suffer from heart problems later in life.

Human nature's paranoia regarding certain risks is an excellent catalyst to fuel *hyper-inflated marketing*. We just finished fighting the cold war, where we were told that we lived in fear of an impending nuclear attack and now these weapons could be in the hands of mad-man dictators or Arab terrorists. They could be deployed in 45 minutes.

Send in the troops.

## Eliminate critical analysis

In Operation Market Garden, *A Bridge Too Far*, during World War II, intelligence officer Major Brian Uruhart discovered aerial photographic evidence that the Nazi's had two Panzer tank divisions resting at Arnhem, only 8 miles from the British airborne parachute landing zone. British top-secret code breakers at Bletchley Park, code named Ultra, had also intercepted secret German coded communication to confirm this. However, no senior official one wanted to hear the truth. They simply wanted to conduct the operation, and as such, the allies suffered over 11,000 causalities. What was the harm in listening to Uruhart? Why at least were anti-tank weapons not issued to the paratroopers? Instead Uruhart was sent on sick leave to silence him.

Once a path is set, critical analysis is viewed as being dangerous to the greater good that started the entire *hyper-inflated marketing* process in the first place. They hold the *party line.*

The process repeats itself again and again. Just ask Hans Blix. When the head of the UN weapons inspection team in Iraq accused the US and the UK Governments of dramatising the threat of weapons of mass destruction from Iraq in order to strengthen the case for war, he was ridiculed. In the end, no such weapons were found.

What is interesting is that each time such an example happens in history the public somehow believe that the case is isolated. A review might be conducted from time to time, but the true facts are only known decades later. Historians might understand, but the public has moved on to its next crisis and the principles are allowed to continue. If the facts were known by the public at the time, history would have been very different.

*Hyper-inflated marketing* has been used outside matters of national security and while highly effective, its impact has been costly to society.

Pottery found in Scandinavia dating back to 3000 BC contained asbestos and its name came from the ancient Greeks meaning *inextinguishable.* The world first saw the health effects from asbestos in the middle ages, when people who worked with asbestos, used extensively to insulate body armour, were dying. I guess that shining armour can kill in more ways than one. Asbestos production stopped until the early 1900s when we started mining it again. In the 1920s, insurance companies started to refuse to insure miners. They knew. Maybe they read history books. By the 1930s, asbestos was listed as a dangerous good and medical publications published its health effects and deaths.

However, the common good took a turn during World War II when asbestos was used in war ship building. The subsequent post war building boom saw asbestos used universally – insulation, plaster, tiles, linoleum, pipes, cladding, shingles, roofing as well as for brake pads and cement. Billed as the *miracle mineral*, Governments and industry turned their backs on the health problems and promoted the benefits for the common good.

The post war building boom of the 1950s made the need for action vital. Inaction meant you might have no place for you and your young family to life. The fire proof nature of the mineral made the wonderful benefits of improved fire safety easy to market.

Industry profited significantly from the use of *hyper-inflated marketing*. Building companies like James Hardie knew the risks. They hired older people to work in their facilities, knowing that the asbestos cancer, mesothelioma, took up to 20 years to develop. It was believed that the older worker would be dead or near death by the time the cancer caught up with them. They forget or didn't care about the young builders who constructed the structures and the rest of us who had to live forever with the threat from the asbestos building materials they used. My wife's father was one of these builders. He contracted mesothelioma at the age of 64, one year short of retirement. After a brave and painful fight, he died on 19 March 2009, aged 68.

Canada, the country that promotes its green credentials to the world, is still mining 180,000 tonnes of asbestos per year and selling 96% of it to the developing world to this very day. Recently, the Canadian Medical Association, the Canadian Public Health Association and the National Specialty Society for Community Medicine demanded that Provincial and Federal Governments stop funding the asbestos industry and promoting Canadian asbestos abroad. The asbestos mine actually receives Government financial support to encourage employment.

Their calls were ignored.

According to the World Health Organization, 90,000 people die each year from asbestos exposure and yet the Canadian Government does not even require exported asbestos to bear a hazardous material label.[86] Principle 4 – eliminate critical analysis.

The common good from cheap building materials was clearly out-weighed by the long term health risks. Abusing *hyper-inflated marketing* principles can have serious consequences.

My next example of *hyper-inflated marketing* happens to be my favourite – *The Y2K bug*. Since this was now 11 years ago, my how time flies, there might be some people who were not aware that when the calendar turned to the year 2000, all of the world's computers were going to stop working. You heard it correct, the software in the world's computers, every bit of it, had to be upgraded to the latest release by a team of highly paid IT auditors. Every company needed to perform a Y2K audit, or they risked being de-registered. For almost 5 years, every ounce of IT resource went into fixing the Y2K bug.

The world's power grids, its telecommunications networks, the banking system, would all be brought to a halt from the Y2K bug.

I went out at 5:30am on the morning of 31 December 1999. Armed with my tent and an esky full of food and champagne, I set up camp for the day in the prime spot on the planet - the shore of Sydney Harbour, right in front of the famous Sydney Harbour Bridge to see the new millennium fireworks display. This show was also computerised. Some people claimed that it may not work because of the bug.

Being the first, I had my choice of locations, yet by 7:00am, the site was filling fast. At the time I was the Operations Director of a telecommunications company. I was given a mobile phone from a competitor's network by the Y2K committee and was told this was to be used in case our network didn't survive the Y2K crisis. I thought to myself, what am I supposed to do if the world does come to an end? Who was I supposed to call? What was I supposed to ask them to do if I did get through?

Of course nothing happened anywhere. The world's banking systems worked, the electricity grid didn't collapse and the telecommunications networks worked. I was able to call my Dad in Pittsburgh at midnight for free from the phone provided. After all, I had to test the network somehow. This might have been the only benefit from the Y2K bug.

Y2K was classic hyper-inflated marketing.

*Common good:* the world's computers need to function;

*Tragic consequences:* the banking, electric and telephone systems will collapse;

*Promote risks to the extreme:* every computer in the world will stop. Every electronic transaction will not work;

*Eliminate Critical Analysis:* I'll offer a personal view here. I was taken off the Y2K committee when I asked why we couldn't just change the date in the computers this weekend and see if it all worked? Worst case is that we would give away some free phone calls over the test weekend. Best case, we didn't have to spend all this money on Y2K. I remember being dragged into a board meeting by the finance director, who chaired the committee, so the chairman of the board could give me the facts on supporting the Y2K audit. "We don't want to be the only company in the world that doesn't conduct this audit." Critical analysis was killed off quickly in the Y2K, as it is in all aspects of *hyper-inflated marketing.*

The easiest way to spot if *hyper-inflated marketing* is being used is to watch how critical analysis is handled. This is a sure give away. If the people who ask focused questions or find evidence contrary to the claims are ridiculed, labelled outside of the mainstream or called sceptics or conspiracy nuts, *hyper-inflated marketing* is somewhere to be found.

So now that brings us to climate change and *hyper-inflated marketing.* Each of us has been subjected to the climate change message to such an extent that it is difficult to make an informed decision. Think about the questions:

- Is it in the planet's best interests for human's to pollute less?
- Are the consequences of continuing to dump $CO_2$ in the atmosphere tragic?
- Are the risks of climate change catastrophic?
- And are climate change sceptics endangering the lives of our children?

Now before you simply answer yes to these questions, remember that using *hyper-inflated marketing* can have possible side effects.

# Chapter 14
## CLIMATE CAUSALITIES

I've been a volunteer member of the ski patrol for the past 7 years. I really enjoy the work – everything from advanced emergency care to providing directions back to the village. It's the first job I ever had where I wear a uniform. You certainly stand out to the public. In May 2009, I received an urgent call from our captain of the ski patrol. We were having our annual pre-season seminar and one of the speakers had pulled out at the last moment. Would it be possible for me to speak again this year? Last year I did a presentation on knee surgery based on my latest marketing assignment at the local private hospital. They were stuck, so I put my head around my marketing and the climate change investigation that I had been doing as a hobby, albeit with gusto, for the last several years. I thought it was a good fit, so I went with it.

It seems that global warming was forecasting the demise of the ski industry and in particular, the Australian ski industry. While the Australian snow fields represent a larger geographic region than Switzerland, the season is very short and the temperatures hover just at the fringe of freezing. Snowfall is unpredictable. With the increases in global temperatures, snowfall is predicted to end as soon as 2020.

Alpine skiing is one of the greatest activities on earth. The feeling of floating down a snow covered mountain, having total control yet bouncing through the bumps at high speed, being at one with the mountain; it is just so special.

The problem is that learning to snow ski is difficult. To become accomplished can take years and like many activities, it is best to start when you are young. Children adapt to snow with such ease. They lack the stiffness and insecurity that adults bring to new ventures. I was lucky enough to be from a skiing family. My first time on skis was at age 2, and by 7, I could ride the poma lift by myself and off I went. No stranger danger in my family. Dad worked for the airlines, so we were able to travel the globe in search of the perfect run. Every weekend during the winter, he would somehow lug us four children up to Seven Springs, our local ski resort near Pittsburgh. He would always have packed warm lunches, including chilli or chicken soup and a great sandwich of chipped ham; a Pittsburgh delicacy of very thin shaved ham.

I reminisce because skiing is not only difficult to learn, it involves great expense in equipment and lift tickets, organisation, winter driving and most important, family commitment.

So now back to the ski patrol seminar. I basically presented what was to become the framework of this book, but with a skiing twist.

You see; if global warming is real than it is going to end the snow and we will have no skiing.

If global warming does not occur, people will have failed to make the considerable investment in learning to ski. No skiers, no ski industry.

Either way, skiers are going to suffer from climate change. Why would you learn to ski if there is going to be no snow? Like the forecast of the polar bear's demise, skiing could be the first activity to become extinct from climate change.

Up until this realisation, I didn't care really care about climate change and the claims in the media. Is it in the planet's best interest for us to pollute less? Of course it is. What difference did it make if the media exaggerated the effects of climate change? All that can happen is that we end up with a cleaner planet. This had to be a good thing.

But not for skiing.

Of interest, the demise of the Australian ski industry has been forecast without even one temperature measurement site for the Australian snow fields contained in the Australian Climate Change Dataset. This is an alpine region with a unique ecosystem that we are told is hanging by a thread because of climate change. Yet not one location is in the dataset. The funny thing here is that I know they take the temperature, as I have done it myself. On the ski patrol, you'll get a radio call to "swing by and check the weather." The fact that the weather station is at the bottom of the hill near the little used and aging Snowgums lift, the term "swing" is a little generous. But I now plan to be *animus of observation.*

But where does this information go? Why is it not in the dataset? To such questions I have no answers but the simple fact that conclusions are made without data however, is clearly wrong.

There is some data for Thredbo on the BOM's general web page. It contains about 85% of the days since 1969. It suggests that Thredbo's maximum temperature has changed little over the years. But many key winter days are missing, so a conclusion regarding the coldness to generate snow is not valid. In fact, there are weather stations all over the mountain. They control the new automated snowmaking. Why this data is not plugged into the BOM's climate records is another good question.

The natural snowfall levels do get measured however. Snowy Hydro, the operators of the massive collection of dams that control the melting water flows from the snowfields have collected the snow depths since 1954.

I was very lucky last season in 2010. My wife decided to take my daughter to the USA to play competitive tennis. This gave me a leave pass to ski for almost the entire month of August.

Up until August, the season was poor. But upon my arrival, it began to snow. And the temperatures were cold. The snow was light and fluffy. We skied down runs that have been closed for years. The snow lasted so well, the season closing was extended by a week.

But 2010 was nothing on our best season in 1964. I was just 3 and living in Pittsburgh, but Snowy Hydro captured the results. The worst season was 1973. See Figure 119.

The trouble with this data is the more you analyse it, the more confusing it is to predict the seasons. You see, over the past 56 years, the snowfall has been anything but consistent.

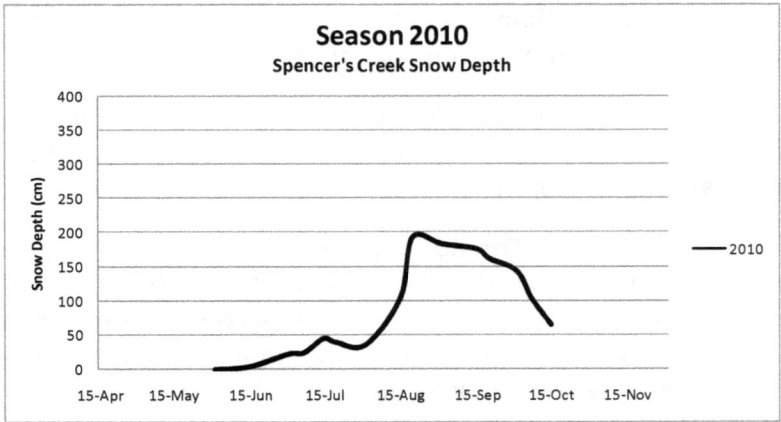

**Figure 118 - Ski season 2010. The snow in August was simply fantastic. Light, fluffy powder. It was called the best season in a decade.**

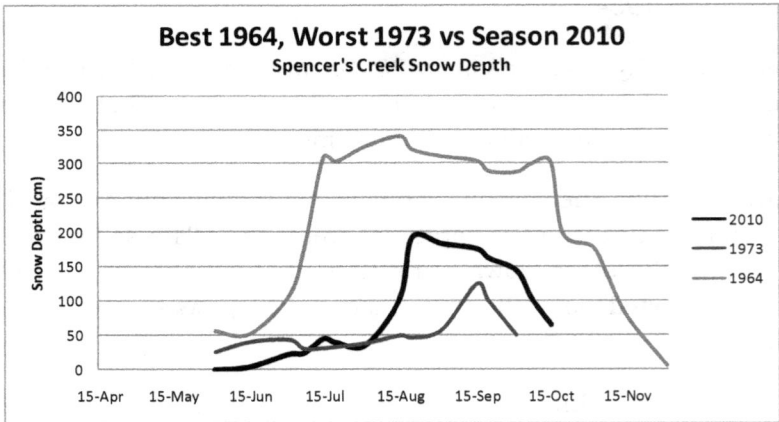

**Figure 119 - Season 2010 was nothing compared to 1964. Season 1973 was our worst.**

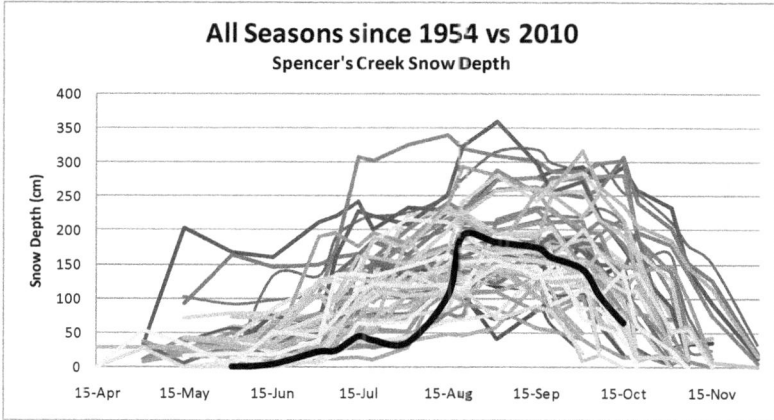

**Figure 120 - Snow seasons are all very different. 2010 was only a good season if you skied in August. If you skied in July, it was one of our worst. Season 2010 is the dark black line.**

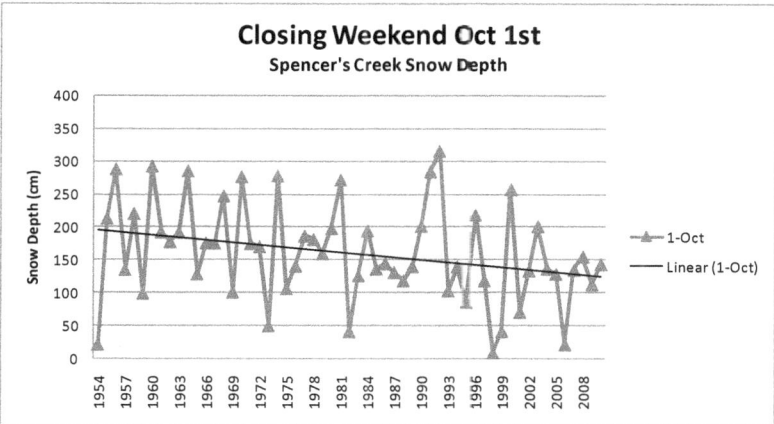

**Figure 121 - The CSIRO claim that snow depths are decreasing by placing a linear trend line on the Snowy Hydro snow depth for the closing weekend.**

173

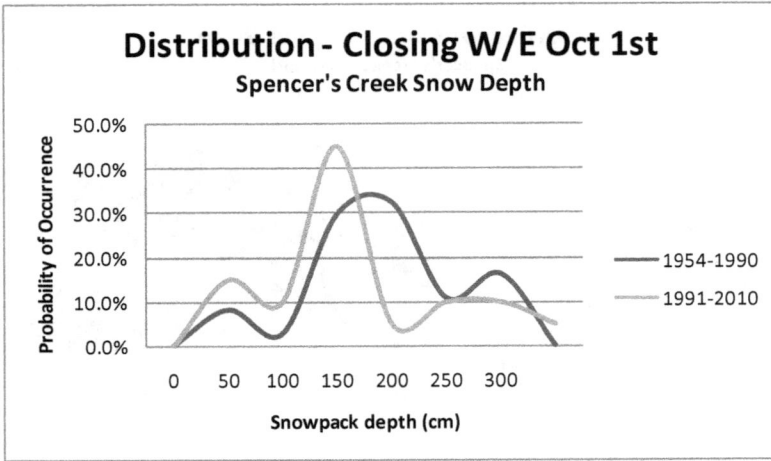

**Figure 122** -**Examining the statistical distribution of the snowpack, there has always been snow on the closing weekend. Today, there is over a 40% chance of 150cm of snow using the 1991-10 distribution. There was a stronger likelihood of having 200cm in the past, but the same likelihood of having 250cm in both distributions.**

The forecast demise of the ski season came from analysis done by the CSIRO.[87] They took the measurements of snow depth on the closing weekend and placed a linear trend line on the graph, showing that the snowpack is reducing. Refer Figure 121. While skiers like to go downhill, the snow pack reducing is not a good thing. The trouble again, does a simple linear trend line really suggest future snow events?

Taking the detailed snowpack measurements for each day, and placing these into a statistical distribution, the trend in snowfall becomes more relevant. Refer to Figure 122.

A few very interesting things about the snowpack come to light when examining the distributions. The early season snow is very similar. By July 1st, there has always been snow. The distribution from 1954-90 and 1991-10 is very similar. The best skiing day of the year has been August 15th. There has been over a 60% chance of snow depth in excess of 150cm.

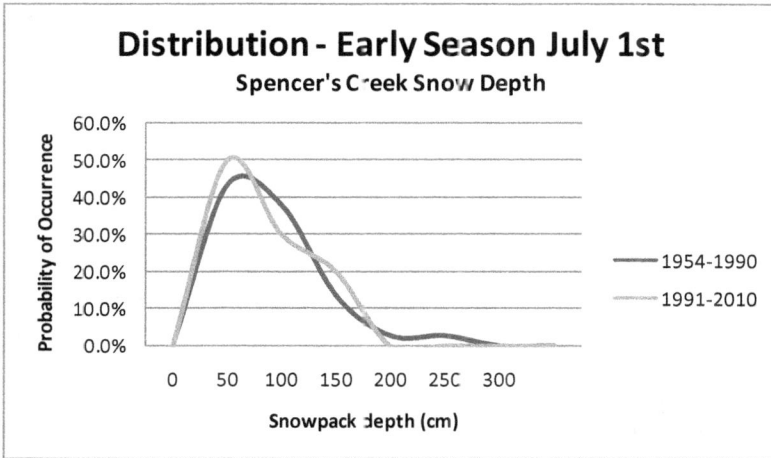

**Figure 123** - The trend in snow is almost identical early in the season. There has always been snow by July 1st.

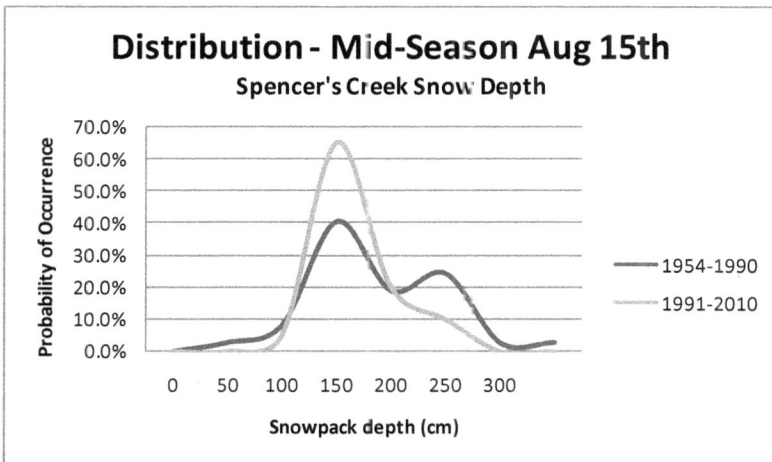

**Figure 124** - Statistically, the best ski day of the year is August 15th. There has been a 60% chance of snow in excess of 150cm.

The closing weekend has also had snow. This time of the year is more traditionally utilised for snow play, as the serious skiers have booked their overseas trips once the snow starts to melt in September.

**Figure 125 - This is a photo of my son Eli, age 5, on the closing weekend in October 2010. Careful management allowed skiing to the bottom of Thredbo – just. How many children will take up skiing in light of the threats from global warming?**

In 1986, I took my first ski trip to the Australian snowfields. Reports were that the snow was excellent. As I drove into Canberra, the nation's capital, I tried to purchase an ice scrapper from the petrol station. The attendant had no idea what I was asking for. You would never drive into the snow in Pittsburgh without an ice scrapper I thought. As I drove into Cooma, the gateway to the alpine region, there was no snow. Another hour of diving I entered Jindabyne – again no snow. From here, I started up the mountain. It was dark. I expected to see snow at any time. I did see some kangaroos and a wombat. There were lots of signs telling me that I might have to put on chains. Again, no snow. I parked my car at the base of the mountain. I got out but in the dark, I still couldn't see any snow. I tucked myself into my sleeping bag laid out in the back of the station wagon. When morning arrived, I woke up and saw the snow line. It started about half way up the mountain.

At the end of the day, pride kept me from riding the lift to the bottom. I walked down through the mud, wondering if top to bottom skiing was ever possible.

Today Thredbo has top to bottom snowmaking. Fully computerised, once the temperature falls below freezing, the snow guns kick into action. The skiing is more consistent now than it ever was. A message the ski industry needs to promote if it is to avoid being a climate change casualty.

But this is difficult to do in the face of *hyper-inflated marketing* of climate change.

Another casualty of climate change is difficult to discuss.

The city of Brisbane is built on a flood plain. We know this well. In the initial inspection of the site in 1824 and 1825, evidence was found that suggested a flood of over 100 feet. Major Edmund Lockyer commented that "marks of drift grass and pieces of wood washed up on the sides of the banks and up into the branches of the trees, marked the flood to rise here of one hundred feet".[88]

In 1841, the highest official level of flood waters was recorded in Brisbane at 8.43 meters. Between 1841 and 1893, the river flooded 22 times. The 1893 flood, while 7 cm lower than the flood of 1841, caused considerable damage, sweeping away the two bridges that spanned the river at the time and leading to considerable loss of life.

The same site as examined in 1824, flooded to 94 feet 10.5 inches in 1893. Maybe we should have taken heed to Major Lockyer's observations. Two major ships, the Elamang and the gunboat Paluma were carried and left aground in the Brisbane Botanical Gardens. Three separate floods actually hit Brisbane in 1893. Nine days after the first flood, a second minor flood only reached a peak of 3.29 meters. A week later, a third major flood carried the stranded ships back into the Brisbane River.

**Figure 126 - Brisbane floods in 1893 with residents rowing around Queen Street. Source: Public Domain.**

Flooding continued into the next century, with a major flood again in 1908. There was a moderate flood in 1931, with 1300 homes drenched. But only a few minor floods occurred in the decades that followed. People and building standards became flood complacent.

Then in 1974, a major flood again hit Brisbane. Not as big as the record floods of the 1800s, this one was still huge. Previous heavy rains made the ground saturated and Cyclone Wanda, while very weak in cyclone terms, was the final straw. The city gauge peaked at 5.5 meters and 6700 homes were flooded.

Following the 1974 floods, the Bureau of Meteorology issued its report into the causes of the flooding and future flood mitigation for Brisbane.[89] It found that there was geological evidence of water levels 5.5 meters higher than the 1974 flood. Meteorological studies suggested that rainfalls well in excess of those experienced in the floods of 1893 and 1974 are possible. The report called for mitigation and weather radar:

Therefore it seems certain that unless major flood mitigation schemes, such as the proposed Wivenhoe Dam, are implemented, floods even greater than those of 1974 will again be experienced in Brisbane.

Development of the weather-watching radar facility with a capability for operational determination of rainfall intensity is considered to be the only way by which the provision of an adequate quantitative flood prediction service for the Brisbane metropolitan creeks is feasible.

The Wivenhoe Dam was constructed. The weather radar was built.

Jump ahead to 2011. The rainfall experienced in South East Queensland was considerable, but not unique. Certainly higher rainfalls could have occurred. The 1974 report called for the possibility of even more significant rainfall events.

On 11 January, Hubert Chanson, a professor in hydraulic engineering at the University of Queensland, told the ABC news that "the dam was built to withstand an event similar to what we are seeing – we have been spared of any flooding in Brisbane thanks to the Wivenhoe reservoir."[90]

The Wivenhoe Dam has a capacity of 225%. Over this level the dam would be breached and flowing water could undermine its foundation, putting the security of the entire dam under question. On 11 January, the dam was already at 173%.

Professor Chanson was optimistic. The Wivenhoe Daw was easily built to withstand the rainfall, but not in the manner in which it was managed.

The same Bureau of Meteorology report from 1974 highlighted the problem of dam management:

There are considerable problems in deciding when to empty the flood storage. If floodwaters were retained by the dam for too long not only would there be major and prolonged flooding upstream from the storage but the dam would become virtually useless for flood mitigation downstream in the event of a repetition of excessive rainfall. Meteorologically such a situation has already occurred (in 1893 when there were three floods within a month) and a recurrence appears inevitable

On 13 January, the Brisbane Courier Mail newspaper reported that the Queensland Police investigated a false text message that was circulated claiming the Wivenhoe Dam wall was breached.

"Dam managers say all dams are safe and operating within design specifications," the newspaper claimed.

This was not true.

The Brisbane Mayor, Campbell Newman warned that the dam could no longer protect the Queensland capital. "The dam is full. Every bit of rain that falls on the catchment can get to Brisbane, and there is not much more we can do about that."

The river peaked at 5:00am on 13 January at 4.46 meters, well below the initial forecast of 5.5 meters. Not as big as 1974 and about half the size of 1841. Yet the damage was more significant, with 33,701 properties flooded.

The Wivenhoe Dam peaked at 190%.

Had a second or third rain event occurred, such as in 1893, the police investigation into the false text message might not have been necessary.

On 21 January, the Australian newspaper reported that leaked emails from the Wivenhoe Dam's engineering office revealed that its operators held on to water for too long. Brisbane's flood could have been largely avoided if action was taken earlier.[91]

The Queensland State Premier, Anna Bligh, has called a commission of inquiry to investigate if the dam's release strategy and subsequent flood was avoidable. It is due to complete its findings by 17 January 2012.

So how does this story become a climate change casualty?

Not to pre-judge the results of the Government inquiry, but the hyper-inflated claims of climate change causing long term, prolonged drought in Australia could easily have had a bearing on the decision to keep the dam close to its maximum capacity.

On 10-11 February 2007, the Brisbane Courier-Mail ran a feature on the water crisis facing Southeast Queensland.[92] The headline read:

# *Bring us a monsoon*
### *Near-tropical storms needed to fill storages*

Rob Drury, the Seqwater operations manager for Wivenhoe and the smaller Somerset and North Pine dams, said "you do need large, uncommon events to fill large dams. You don't fill them every year." He told us that Wivenhoe has the capacity to store 1,165,000 megalitres as well as an additional capacity of 1,450,000 megalitres to mitigate flooding. He added, "There have only been four main rainfall events in the past 15-16 years. It has been seven years since we had a major rainfall event that has given us a refill of 50 per cent of the dam."

Seqwater is South East Queensland's bulk water supply provider. Taken from their corporate website:

> We deliver innovative and efficient management of catchments, water storages, and treatment services to ensure the quantity and quality of the region's water supply.

Actually, Seqwater is their trading name. Their actual name is Queensland Bulk Water Supply Authority. They were established in 2007 as part of a water reform agenda by the Labor Queensland Government under Anna Bligh - the same Anna Bligh who called for the enquiry into the dam's release strategy. Her website lists water reform as one of her main achievements.

Seqwater's *Strategic Plan* is available from their website.[93] Their first strategic goal is to supply customers with reliable water of quality. They took great pride that a new, single focused organisation was created out of 14 regional water entities. Their new mission statement is all about catchments:

> Catchments are vital regional resources. We define catchments as the combined natural and built infrastructure needed to source, store and supply water to meet the quality and reliability needs of our customers.

In the 36 page *Strategic Plan*, flood mitigation gets mentioned only 3 times. One of these references is actually linked to green energy and recreation.

Their number one key performance indicator in *whole-of-catchment-knowhow* is: **budget achievement** – "the degree of accuracy with budget forecasts."

Did someone forget to tell them that the Wivenhoe Dam was actually built to flood proof Brisbane? A city built on a flood plain. A dam built after the 1974 public outcry.

Southeast Queensland is the fastest growing region in Australia. The Queensland Department of Infrastructure and Planning estimates that the population will continue to grow from 2.8 to 4.4 million people by 2031.

To handle this growth, development permits allowed housing and businesses to be established in known flood prone areas. So now a flood of half the size impacts double the amount of people.

Climate change hype told us that drought was our fear. We were told that rainfall patterns have been changed forever. And the proud Queenslander, Kevin Rudd, told us there was nothing worse than being a sceptic. You were not allowed to question these claims.

So Seqwater's Rob Drury prayed for a monsoon. And he got one! 631mm of rain fell in Lindfield, part of the catchment, in the month of January, a massive 257mm on 10 January alone. With this water flowing into the dam, and its level already at 173%, the cities of Brisbane and Ipswich were doomed.

The actual weather event was quite isolated. At the Cape Moreton Lighthouse, near Brisbane, just 26.2mm of rain was recorded on 10 January 2011. But it was not a surprise, the rain event was watched on the weather radar that was supposed to provide additional safety.

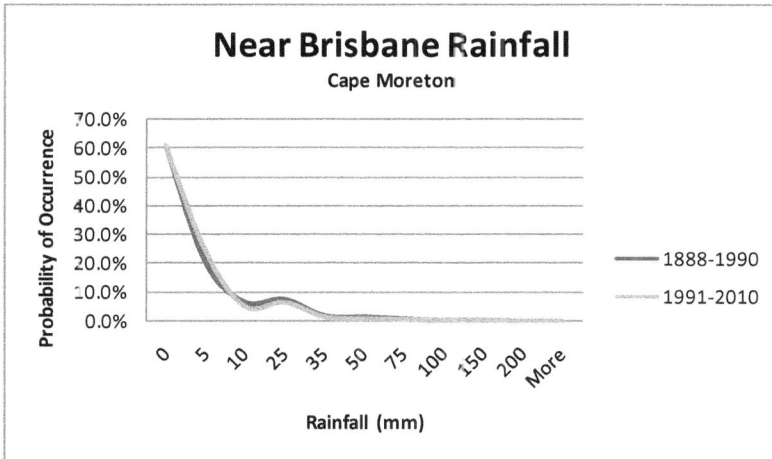

**Figure 127 - The rain profile for Cape Moreton, near Brisbane has not changed since 1888. There is a 1.34% change of rainfall over 75mm. The dam was built to mitigate this flood risk in Brisbane.**

Extreme weather events happen in South East Queensland. Even with the feared drought, the rain profile has not changed in over 100 years. From 1888 to 1990, there was a 60.8% change of it not raining. From 1990-2010, the percentage was 60.9%.

The really ironic thing is that we don't learn from our mistakes. Queensland has a history of repeat flooding. On 3 February, Tropical Cyclone Yasi crossed the north Queensland coast. In 1974, the floods worsened when a weak cyclone named Wanda dumped 526.5mm of rain on Brisbane over two days. Had Yasi, a Category 5 storm, travelled down the coast rather than proceeding inland, Brisbane would have been inundated with rainfall again, just like the second flooding in 1893. The Wivenhoe Dam operators again ignored this threat. The dam was 100% full in early February. Realising the risk, on 14 February plans to reduce the dam levels to 75% were announced. The media reported that the 291,000 megalitres of water to be released represents a one year supply of drinking water for Brisbane.

It is obvious that doubling the population requires more water. Queenslander's had their chance to resolve this problem when a dam across the Mary River in Traveston was proposed to solve the water crisis in 1996.

The dam came under strong local protests, but it was the environmental concerns over several fish and aquatic species believed to be under threat from the dam that received the greatest attention. The most significant was the Mary River Turtle, who breaths through its bum and the Queensland lungfish. In the end, climate change won the battle – no dam. It was calculated that rotting vegetation from the shallow waters would generate more greenhouse gas, in the form of methane, than the $CO_2$ needed for sea water desalination. In November 2009, Labor's Federal Environment Minister, Peter Garrett, refused approval for the project.

As reported in the Brisbane Times, Australian Greens Leader Bob Brown congratulated Peter Garrett on his decision to reject Queensland's controversial $1.8 billion Traveston Crossing Dam - "I've sent a letter of congratulations to minister Garrett on making the only decision an environment minister could have made, that is, to veto the dam," Senator Brown said.

Losses and rebuilding costs for the Queensland floods are forecast to be in excess of $16.78 billion.

The impact on the ski industry and the failure to perform flood mitigation are just two examples of *hyper-inflated marketing* for the common good that resulted in unfavourable outcomes.

# Chapter 15
## CARBON TAX

Coal is plentiful. Coal is dirty. Coal is cheap. Coal needs to be taxed.

The heat from burning it drives the turbines and produces electricity. A by-product of burning coal is $CO_2$. The other by-product is old fashioned pollution. Coal is very dirty.

Our roads are also full of cars and trucks that produce even more $CO_2$. Yet we are told that this problem can be solved with hydrogen.

Burning hydrogen in vehicles has been talked about for over a century; Hydrogen fuel cells were discovered back in 1839 by the Welshman, Sir William Grove. Burning hydrogen is very clean because its only by-product is water. The trouble, however, is not with the burning part it's with the making part. Where do you get the hydrogen to put in the cars?

Most of us have performed this experiment in chemistry class. You use electrolysis, with a metal probe connected to a battery, to break hydrogen and oxygen from its natural state of water ($H_2O$). Water is not the best source of hydrogen, electrolysis of methane ($CH_4$), from natural gas, is much more commercially viable.

Of course, electrolysis is defined as the process of using a direct electric current to drive an otherwise non-spontaneous reaction. In other words, water and methane just don't break apart without some help. We need electricity. Oops, we are right back to that dirty coal again.

To address the problem of climate change, Governments around the world turn to what they do best: Tax.

In Australia, the Government commissioned Professor Ross Garnaut, one of its most distinguished and well-known economists, to develop an economic plan to tackle climate change. His review only examined the economic risks of climate change. The scientific proof came from the IPCC report, as summarised in the Garnaut report:

> The Review is not in a position to independently evaluate the considerable body of scientific knowledge; it takes as a starting point the majority opinion of the Australian and international scientific communities that human-induced climate change is happening, will intensify if greenhouse gas emissions continue to increase, and could impose large costs on human civilisation.

China, and other developing countries, is his major concern. China has now overtaken the United States as the world's largest emitter. Without mitigation, "developing countries will account for 90% of emissions growth over the next two decades, and beyond," his report said.[94] He called for the development of an Emissions Trading Scheme, centred on the price of coal. His report makes this very clear:

> High petroleum prices will not necessarily slow emissions growth for many decades because of the ample availability of large resources of high-emissions fossil fuel alternatives, notably coal.

Call it what you want - an *Emissions Trading Scheme, Carbon Pollution Reduction Scheme, Carbon Offset, Carbon Credits* – the objective is the same, tax carbon to make coal more expensive and encourage alternative forms of energy to generate electricity.

Of course, there are other ways to make electricity other than burning coal.

Hydro power, where the water flowing out of the dam drives the turbines, is viewed as being very green energy. Except during the damming part, where the flooding water kills off eco systems both up and down stream. Building new dams is always an environmental challenge.

There are no shortages of dams in China. The Yellow River, China's second largest, has 13 major hydroelectric power stations. Across its largest river, the mighty Yangtze, is the Three Gorges Dam. This is the world's largest hydroelectric power station, with the capacity to generate an incredible 100 trillion watts of electricity per annum.

The Chinese had to relocate 1,240,000 residents to make way for the dammed water. Could you imagine the political fallout in building such a dam in the West?

With such a herculean effort, no one could say that China is not trying to be green.

But don't get too excited about its use of green energy. China is still the world's largest consumer of coal, generating almost 70% of its electricity needs; the US generates about 50% of its needs from coal. China has a lot of coal. It only sits behind the US and Russia with the largest coal reserves. And there is no shortage of countries wanting to export coal to China; in Australia, 75% of all the coal mined is exported.

Nuclear power is very green. A controlled radioactive reaction creates heat which drives the turbine. There is no $CO_2$ produced in a nuclear power plant. The large towers that are symbolic of the nuclear industry are water cooling towers; where the steam that was made to turn the turbine is allowed to condense and trickle down the towers to cool in the air. The 'smoke' from these towers is simply water vapour. Of course, while nuclear is very green, there is always the concern of a nuclear accident such as the Japanese crisis filling today's headlines or the accidents that occurred in 1979 at Three Mile Island in Pennsylvania or in 1986 at Chernobyl, in the Ukraine. And did I mention what to do with the nuclear waste?

I remember the accident at Three Mile Island very well. Our next door neighbour worked at Westinghouse, designing nuclear reactors for navy submarines. We only found this out one day when a letter was wrongly delivered to us and the address had "Dr" in front of his name. I was told that most of his work was classified so we never really talked about it much. But he was asked to review the situation at Three Mile Island as the leakage was very similar in nature to that experienced in a reactor failure in a submarine. We had lots of Government cars parked in front of the house during the crisis. At the time, it was very exciting.

When I was at Penn State we used to ski at Roundtop, a small resort near Three Mile Island. The joke used to be that you didn't need lights for night skiing because everything glowed from the radiation. This wasn't true. In 1983, I flew my Cessna over the plant. I had a good look down inside the cooling tower. I'm sure the air space around a nuclear power plant is now closed, so this was again something very few people will experience. I have to report that there were no glowing green mutants.

Three Mile Island's other reactor was reopened and operates to this day, producing 802 megawatts of power. In 2009, its operating license was extended to 19 April 2034. After 30 years in storage, the generator connected to the damaged reactor has been refurbished and shipped to Shearon Harris nuclear centre in North Carolina.

Of course, the situation at Three Mile Island was nothing compared to the explosion at Chernobyl in 1986. A reactor test went horribly wrong and resulted in an explosion that released large amounts of radioactivity and killed 70 workers at the time. Acute radiation syndrome (ARS) was diagnosed in 134 emergency workers, 28 of which died in 1986. Since then, 19 of these workers have passed away, but their deaths may not have been related to ARS.

Radioactive particles were measured as far away as Scotland. The reactor was sealed in a concrete sarcophagus, which is now being redone because it has leaked. An exclusion area around the plant was established in parts of Ukraine and Belarus.

At the time, the media reported that the accident will cause tens of thousands of cancer related deaths. While no one would desire to be exposed to radiation, aside from an increase in childhood thyroid cancer, the World Health Organization found no evidence exists to suggest that other cancers, including leukaemia, have increased in the population effected by Chernobyl.[95] Their extensive study also found no increase in birth defects, such as Down's syndrome or increased infant mortality. There was also no evidence of decreased fertility among males or females.

The World Health Organization found this intriguing as epidemiological studies pointed to different results. Their report said "no increased risk of leukaemia linked to ionizing radiation has so far been confirmed in children, in recovery operation workers, or in the general population of the former Soviet Union or other areas with

measureable amounts of contamination from the Chernobyl accident."

Since the incident released large amounts of radioactive iodine (io-
dine-131), children were particularly vulnerable. More than 4000
cases of thyroid cancer were diagnosed in Belarus, Russia and
Ukraine between 1992 and 2002, most of these attributed to radiation. The majority of these patients were treated successfully. It
was found that the distribution of iodine tablets and evacuation
helped to minimise the consequences of the accident.

While the exclusion zone around Chernobyl is devoid of humans, it
is teeming with life. The zone has become a nature sanctuary
where species that haven't been sighted in the region for decades,
like the lynx and eagle owl, have returned. Birds are even nesting
inside the sarcophagus. It seems the animals prefer plutonium to
pesticides, industry, traffic and drained marshlands. Large mammals, like the wild boar, bison, elk and wolfs freely roam in the
zone. Plants are even regrowing in the *Red Forest*, an area initially
killed by the highest levels of radioactivity. Evidence has been
found of DNA mutations in some animals, but nothing that affected
the animal's physiology, life span or reproductive ability.

The Centre for Disease Control and Prevention found more than
10,000 US miners have died from black lung in the last 10 years.
Clearly, coal has killed more people than the nuclear power accidents at Chernobyl and Three Mile Island.

Hopefully the release of radioactive materials in Japan has minimal
long term health effects. As this crisis was unfolding at the time of
publication, comments on the specific events and actions are not
possible. What is clear however is that prior to the earthquake and
tsunami, Japan was committed to increasing its nuclear energy capacity. The only country to receive a nuclear attack, today the
Japanese have 55 nuclear plants and are in the process of building
11 additional reactors in 7 plants. The Fukushima Daiichi plant,
northeast of Tokyo has the six damaged reactors. Its first reactor
commenced operation in 1971. While Japan has not had a magnitude 8.9 earthquake since AD869, the lessons learned from the
2004 Asian Tsunami show us that even far away seismic events
can cause major destruction. Massive earthquakes are common in

the Pacific. For example, in Chile, two huge earthquakes struck with a magnitude 9.5 in 1960 and a magnitude 9 in 1869; a magnitude 9.2 quake rattled Alaska in 1964 and closer to Japan, a magnitude 9 quake hit Kamchatka in 1952. These seismic events could easily cause severe damage in Japan. Yet for Tokyo, this wasn't the big one. Its fault line last moved in 1923, with a magnitude 8, killing over 140,000 people in one of the worst disasters of the 20th century. Yet Christchurch, New Zealand shows us that the magnitude is not as important as the location and depth. The town easily survived a magnitude 7.1 event in September 2010, but a 6.3 magnitude quake with its epicentre in a worse location caused wide spread damage and death. No matter how clever your engineering is, building nuclear reactors in known earthquake zones is not a good idea.

Yet power is needed where the people live – principle of common good. The population of greater Tokyo is over 12 million and its Government has used *hyper-inflated marketing* techniques to ignore the lessons learned from the Asian Tsunami and warnings over 2 years ago from the International Atomic Energy Agency (IAEA) that a strong earthquake could pose a "serious problem" for its nuclear power stations, Britain's Daily Telegraph reported.[96]

It seems the Fukushima Daiichi plant was only designed to withstand a magnitude 7 earthquake. I guess this is now obvious, since the plant clearly didn't survive the event. Since the Richter scale is logarithmic, meaning that whole-number jumps indicate a tenfold increase in amplitude, there is a 100 times difference between the design point of a magnitude 7 and the magnitude 9 earthquake that rocked Japan.

Even if the events in Japan are successfully contained, the image of nuclear energy has been dramatically damaged. For example, in Germany, 40,000 protestors recently formed a 45 kilometre human chain in protest of the German Government's decision to extend one of its 17 nuclear power plant's operating life.

The United States has 104 reactors, 52 of which have been operating for over 30 years. Most of these operate in the geologically stable east coast. However, a couple major facilities do operate in earthquake prone California. Again, power is needed where the people live.  China has 7 nuclear power plants with plans to build more.  Today these are located in areas of low seismic activity.  The capital Beijing, with its 22 million people, sits on a zone of moderate to high seismic hazard.  This is another reason why coal might represent the only viable alternative for a seismically active country like China.

In Australia, a country rich in uranium with ample sea water for cooling and virtually no earth quake risk, the interest in nuclear power is non-existent.  It is simply not popular.  Australians prefer to export their nuclear fuel reserves, burn coal and look to alternative green energy sources.

This brings us to wind and solar power.  Both these technologies are billed as the answer to our energy and climate needs because neither of these technologies produces any $CO_2$.

Individually, windmills are huge - larger than the wingspan of a Boeing 747.  Grouped together, they are called *Wind Farms*, a romantic name for a massive hunk of steel bolted into a reinforced concrete foundation situated on a rounded hilltop.

Ideally, they have to be stationed in a zone with continuous, steady winds.  If the winds are too strong, the mill is shut down to avoid damage to its internal gear box and generator.  On windless days, the mill does not generate electricity at all.  Most wind farms are placed in a picturesque location in a windswept valley or on the hill line.  Quite a few people are concerned about the visual pollution of the wind farms on a mountain top or along the coast.  The propeller, while it looks like it is spinning slowly at a distance, actually reaches speeds at over 160km per hour at its tip.  This can kill birds and bats - another environmental concern for some people.  What is certain, with all the moving parts, windmills require considerable ongoing maintenance.

Since most people have chosen not to live in windy places, the use of home sized wind mills, while available, are not really going to tilt the balance away from $CO_2$.  But solar comes in two varieties - massive and domestic.

Solar technology uses a photo cell, constructed out of glass and silicon, to collect the sun's rays and convert it to electricity. It doesn't work from the sun's heat, rather it uses the energy from the light rays; but bright sunny days with large amounts of sunshine are very useful to increase yield. There are some solar technologies that utilise heat, such as hot water systems or swimming pools, and commercial units where superheated oil can be used to drive a turbine. But the base solar system only needs sunlight.

While there is a great deal of hope and public debate regarding these technologies, when you boil it down, wind and solar suffer from the same two problems.

1. The electricity they produce is unpredictable. No sun, no wind, no juice. As such, they always need to be combined with another form of generation or storage system to produce electricity on demand. At the risk of stating the obvious, solar power does not work at night.

2. Windy and sunny places, with ample available space to install solar panels and windmills are not generally situated near cities or on the power grid. As such, major new transmission systems need to be constructed and funded.

In addition, coal generation does not compliment well with these green systems. When you burn coal it needs to be hot - the fire needs to be roaring. You can't just turn it down some to make less electricity. If the temperature gets too low, the coal doesn't burn well. So to make the coal plant efficient, it needs to run continuously. There has been a great deal of criticism that wind power has not replaced one coal power station. The reason for this is simple. Wind and solar cannot be the sole source of power.

These technologies can be supplemented by gas powered stations. Gas can be turned up and down quickly as it is highly combustible. Yet burning gas generates $CO_2$ so these stations are regarded in the same category as coal. This might not be fair for the gas industry, but the green/gas combinations have not been promoted and thoroughly analysed in the public domain.

Wind and solar farms also need to have their electricity transported from the generation site to the grid. This can more than double the cost of these technologies. Trying to store electricity is also very expensive. A farm of batteries to collect solar during the day and use it at night is really expensive and has very high on-going maintenance. Linking a wind and solar system to a dam, re-pumping the expelled water back into the dam for use in peak periods might be a solution, but in practical terms, this concept is not going to solve the global warming issues on the planet. It cannot be done on a large enough scale.

Let's take a moment to understand the economics of the electricity industry by analysing an electric bill. The power industry claims that the cost of the actual generated electricity is only 40% of the total. The largest component, 46%, is the cost to develop and maintain the transmission lines. The other 14% covers the retail costs including invoicing, customer support and a retail profit margin.

Electricity can be generated by the power company or purchased from the generators connected to a national grid at a wholesale rate. According to the Australian Energy Market Operator (AEMO), the wholesale cost of electricity in my home state of New South Wales was $22.70 per megawatt hour on 29 June 2009. This works out to 2.3c per kilowatt hour. We are charged 19c per kilowatt hour by Integral Energy, the electricity company that supplies power to our home. At the 40% electricity component, this means that we are charged 8c per kilowatt hour for the power and 9c per kilowatt hour for the transmission.

Now we are all told that Australia's electric companies, because of market uncertainties – most notability a future carbon tax - have failed to make the necessary investments in upgrades and maintenance of the power transmission lines. So regardless of the price of coal increasing from a carbon tax, the consumer will pay more for electricity. This seems to be a common argument used globally by power companies. More money is needed for infrastructure.

I'm sure it has nothing to do with profiteering and a general move from public to private ownership in power utilities.

That brings us now to the generation?  Coal generation obviously costs less than 2.3c per kilowatt hour, the wholesale price paid to the generator.  The Loy Yang power station, in the Latrobe Valley near Melbourne, is Australia's largest open cut brown coal mine.  Brown coal has higher water content than black coal.  As such, it is more difficult to burn and it makes more pollution.  Loy Yang is massive.  It consumes approximately 60,000 tonnes of brown coal per day, transported on a special conveyor system straight from the mine into the power station.  It is dried, pulverised and burned to create high pressure steam at 1300°C.

So what does it cost to make the wholesale electricity that sells for 2.3c per kilowatt hour?  The generator makes a profit.  The Government makes a royalty on the mining revenue.  Based on my calculations, the cost of the coal to generate electricity in Australia is about 1c per kilowatt hour and as low as 0.25c in China.

Self generation – this must be the answer.  Install a solar panel on your roof and the global warming crisis is solved?  The Australian utility companies are promoting and discounting 1.5 megawatt solar systems.  These systems consist of 8 panels and an inverter, which converts the direct current (DC) from the solar system to alternating current (AC) used in the home.  The cost is about $3000.  I know the cost of solar panels is coming down over time, but let's use this price for the example.

Now while the average home in NSW doesn't often use more than 1.5 kilowatts in an hour, the solar system doesn't generate that much power either.  This figure represents the rated maximum power of the unit.  The actual power generated by a 1.5 kilowatt system during a summer's day is around 7.4 kilowatt hours.  In the winter, this is reduced, because of less hours of sunlight, to 5.3 kilowatt hours.  The average home in NSW consumes 20.5 kilowatt hours in the summer and 26.7 kilowatt hours in the winter.

In the summer, the solar unit meets the needs of the household for about 4 hours per day.  In winter, the unit never provides adequate power.[97]

I'm not sure that roof space will be large enough to fit 28 solar panels.  But at 185 watts each, this will rate at 5.18 kilowatts and provide a daily summer profile of 28.8 kilowatt hours but still only 20.8 kilowatt hours in the winter. Some cleaver use of electricity might get you through those winter months. Today, this system will set you back $19,995. Again, let's say that the unit halves in price and we can find space on the roof.

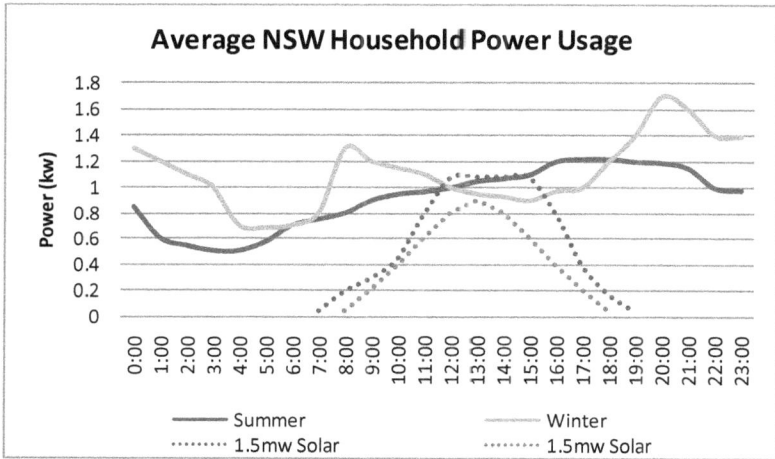

**Average NSW Household Power Usage**

Figure 128 - Average NSW household power usage during the summer and winter, with the output of a 1.5 kw solar system. It does not provide adequate power for the household's needs.

Can we eliminate the transmission network? No. We still need it to provide electricity to people who can't put in solar systems and for when the sun doesn't shine. So we still have the transmission component of 9c per kilowatt hour. That leaves us with retail cost of 8c per kilowatt hour for generation. Remember, the wholesale component is only 2.3c which costs the generator 1c.

Again, I'm not sure that our 5.18 kilowatt collection of solar panels will last 7 years and it will need to be cleaned and serviced, costs that have not been factored in to this analysis.

But that Carbon Tax will have to raise the price of coal 16 times to make these systems just break even. Even when factoring the price of the panels to halve. And we didn't even include the thousands of dollars for an acid battery bank if we seek a consistent supply of power during the day and night.

Home generation just doesn't deliver the economies of scale. Transmission systems still need to exist. Commercial wind and solar farms need to be supplemented by gas-fired power. Dams are effective but there just aren't enough rivers without lung fish and unique turtles. The political will to build dams has changed since 1947 when the decision was made to dam the Snowy Mountains.

In 1983, the High Court of Australia, in a 4-3 landmark decision, found that the Federal Government had the authority under environmental grounds from its imposed *World Heritage Act* to stop the construction of the Franklin Dam proposed by the State of Tasmania. This decision effectively stopped dam construction in Australia.

Since Australia has rejected hydro and nuclear power, there is simply no obvious alternative to coal generated electricity.

**Figure 129 – Construction began in 1948 of the Snowy Mountain hydro scheme's 16 major dams, seven power stations, 145 kilometres of interconnected tunnels and 80 kilometres of aqueducts. Viewed as an engineering wonder of the world, the project was completed mainly with migrant labour for only $820 million in 1974. (About $6.5b in today's dollars.) Electricity was first generated in 1955 and the project set many world tunnelling records. This is proof that Australia had the capacity for big vision projects. Photo: Constructing Tumut Pond Dam, 1960, National Achieves, NAA: A1200, L36209.**

Globally, does anyone seriously think a carbon tax imposed in Australia will stop the Chinese from producing $CO_2$?

China is embracing far more alternative energy generation than most of the world. The Chinese are doing much better than Pittsburgh when it held the world's manufacturing mantra. But since coal only costs 0.25c per kilowatt hour in China and they have the world's third largest reserves of coal on the planet, why wouldn't they use it? Nothing can stop China's $CO_2$ production unless we consume less.

Our problem is simply our desire to consume. But if we consume less, the economy will start to slow and jobs will be impacted. A side benefit of the Global Financial Crisis was the slowing in the economy slowed down $CO_2$ but it also almost brought the planet to its financial knees.

Another good question is if the world's Governments do succeed in introducing a carbon tax, what should they do with the money? It would be a total waste of funds to give this money to households so they can generate their own electricity when the economies of scale make this much better done centrally. But local generation is the most visual sign of tackling climate change. So on 17 February 2011, the Australian Greens announced it will support the Government's flood repair levy provided the solar flagships scheme is restored. What you can do with 11.7% of the vote.

Some economists believe that carbon tax can be revenue neutral, lowering other taxes to compensate for the increase in electricity. Since most generation in the world is done by private industry, will the funds from these taxes be given in incentives or as tax rebates to large utilities that are already very profitable? Should the tax pay for more health care services or perhaps top up the insurance industry that will bear the brunt of extreme weather from climate change?

This debate has not occurred. It should happen before any tax is introduced. But in February 2011, Julia Gillard announced that with her Green Party alliance, a carbon tax would be introduced by July 2012. Her climate change economist, Ross Garnaut, has already recommended that a price of $26 per tonne be placed on $CO_2$

emitted, raising $11.5b in its first year alone. Just for emphasis, Ross Garnaut's report did not review any of the science regarding climate change. But you could also argue that in supporting this tax, Garnaut has now made a serious error in economics. Without an international agreement on carbon, the Australian carbon tax is simply a tax on production that will make the country less competitive in the international market. Coal will continue to be exported and $CO_2$ levels will continue to increase.

This brings us finally to the matter of the IPCC and what to do next. The momentum of the climate change process is so significant that it is almost impossible to influence.

On 23 June 2010, the IPCC announced that a panel of 831 "highly qualified" experts have been selected from over 3000 nominations to produce its fifth assessment report, AR5.

The outline has already been agreed:

    Topic 1 – Observed changes and their causes
    Topic 2 – Future climate changes, impacts and risks
    Topic 3 – Adaptation and mitigation measures
    Topic 4 – Transformations and changes in systems

Meeting locations for the working groups are global, including China, France, Morocco, Bangladesh, Brazil, Belize, Japan, India, Peru, Germany, Abu Dhabi, Switzerland, Belgium and the Gold Coast in Australia.

Audit is a vital function.

Jay Lorsch is the Louis E. Kirstein Professor of Human Relations at Harvard Business School. Following the failure of audits to identify the problems looming at Enron, he wrote an opinion piece in the Financial Times which argued for legislation to create an independent, self-regulatory organisation to oversee accounting firms. Enron, he says, "is not an isolated incident." He was right. Such an organisation might have made a difference in the global financial crisis. His opinion transcends the climate debate as well, "over the past five decades, accountants have changed from watchdogs to advocates and salespersons."

In business, we were, and are still, naïve to believe that thousands of well credentialed auditors will do the right thing without a support structure and ability to be heard as whistle blowers.

In climate science, just because 831 highly qualified experts have been chosen by the Intergovernmental Panel on Climate Change does not mean we have any sort of audit function into this massive organisation. How are individual voices heard and how is evidence verified and selected?

The only audit function that has been performed, the panel investigating the Climate Research Unit, found serious issues that should have been addressed. Based on the outline for AR5, not only have they not been addressed but they have been ignored.

Selection onto this expert panel is almost akin to winning a top award. Almost every member is from a University. Locked in a campus environment, can you imagine the excitement when they get their ticket to Morocco and depart for an international climate working conference? How many of them will say we should allocate the resources to fill in the missing temperature records and place these in a central database for the world to analyse? How many of them will call for the world's weather observation stations to stop closing? How many would call for the inclusion of professional statisticians to review the climate models?

And even if they wanted to, how could they when the agenda is already established? They are not investigating climate change, they are trying to find a word better than *unequivocal*.

Like Jay Lorsch's accounting advocates and salespersons, these experts are likely to be just hyper-inflating marketeers.

Good science, following proper methods of experimentation, analysis and the critique of results against a proposed hypothesis, eliminates sceptics. The IPCC and its feed in organisations, such as the WMO, the British Antarctic Survey, the CSIRO and the Climate Research Unit, have created the climate sceptics by following poor scientific method and encouraging *hyper-inflated marketing*.

The poor science supporting climate change will not be fixed by a tax on carbon or writing AR5. *Hyper-inflated marketing* may make such a tax possible, but it will never be popular unless it is supported by factual and audited science.

# Chapter 16
## MAKE IT MAINTAIN IT

I'd like to offer a different perspective and a solution for the problem. *Hyper-inflated marketing* has been used to communicate climate change to the masses because the science doesn't really stack
up but the problem of pollution in general is a global concern. The
first principle of *hyper-inflated marketing* is to be convinced of the
common good.

A planet with less pollution is a better place. This is clearly common good.

Governments use tools like *hyper-inflated marketing* to motivate its
population to action. Just look at how unconcerned we are as a
society. In the US, where voting isn't mandatory, only 56.8% of the
voting population actually bother to vote. You couldn't have had a
more focused election in 2008. Barrack Obama and his agenda of
change versus war hero John McCain and his female Vice President
Sarah Palin. Yet 43% of Americans failed to vote.

Until more of society actively participates, can you blame Governments for using *hyper-inflated marketing*?

In my opinion, the real trouble with global warming is not the use
of bad science but that Governments have the wrong long term solution. A carbon tax will not solve the problem. It will not produce
a cleaner world and it could drive us into economic recession.

The real problem with the planet is that we have become a throw-
away society. Everything we make gets thrown-away. This generates a continuous cycle of manufacturing, producing more and
more pollution.

It bothers me to walk into a Wal-Mart, the world's largest retailer,
and be greeted by a person my age, who used to work in a factory
earning a good wage for their family. At least they have some work.
I knew so many fathers of my generation that got laid off in Pittsburgh and never found work again.

Yet the rich get richer. Industry is profiteering from the low wages paid in China and India and they also gave away all our manufacturing knowhow for peanuts.

I challenge you. Try to find something not manufactured in the developing world today. Even if it is assembled in the West, the parts come from China don't they?

The problem however is not where it is manufactured. The problem is that we now engineer everything to be thrown away as soon as it breaks. We even engineer things to break, just so you have to throw them away and consume them again. A one year product warranty now means to a design engineer to build a product that will break after one year. Gone are the days of high quality products. This cycle of consumption is the enemy of our planet.

Raising the price of coal from a carbon tax will not overcome this problem. We will still consume and as shown in Chapter 15 changing the price of coal is not going to reduce our dependence on carbon in the economy. We have to consume less. Also, for Ross Garnaut's plan to work, he has to get every nation in the world to agree that paying more for electricity is a good thing. And we have to use this money to encourage investment in alternative generation technologies when the alternatives all have serious limitations. This is not going to happen. Also, the developing world is not going to come to the party.

What we need is a model that consumes less but doesn't impact economic growth - *Make it Maintain it.*

We are very good at the making part. But the throw-away society is everywhere. Let's start with my favourite example, bottled water.

Purchasing a 500 ml bottle of water could set you back up to $2.50 in a supermarket, and even more in convenience stores. This is 6870 times more expensive than tap water. Water is often more expensive than a litre of milk, petrol or ice cream.

It's worth repeating, bottled water can cost more than petrol.

The marketing of water is a classic example of *hyper-inflated marketing*. It was found that as a society we needed to consume more water. We were told that doctors recommend we consume 2 litres of water a day. The tragic consequences of not drinking enough water can lead to poor health and even early death. The risks of not drinking bottled water include bad taste and increased exposure to harmful bacteria.

The fact that tap water is often what is put in bottled water and tap water is safe and clean in places like the US, Canada, Europe and Australia was ignored. Research shows that most people purchase bottled water because it was handy and considered a useful bottle that could be refilled. Only 16% thought it tasted better, but blind test tests failed to prove that it tasted any different. Yet we don't really reuse the bottles. In the US alone, over 17 billion barrels of oil is used to create the 30 billion plastic bottles placed in landfills every year. It is estimated that this entire process produces 2.5 billion tons of $CO_2$.

As a society, we have replaced a perfectly adequate and healthy water delivery system with a product that is 6870 times more expensive and requires us to pollute the world to create and dispose of plastic bottles. And this doesn't include the cost to transport and chill the water for sale.

Yet have we been mislead into the health benefits of water? Kidney Health Australia now promotes that, "there is a lack of evidence that drinking water in excess of thirst is beneficial for the health of Australians living in temperate regions and not exercising strenuously."

Bring back the drinking fountain.

But *Make it Maintain it* is also about high technology. I've had to throw away three iPods since 2005. Apple, the company held up as a great innovator, leads the way in the throw-away society. Sure, each new iPod is an advance on the previous model. But it was only the batteries that failed and the screen which lost its contrast. Why can't I simply replace the screen on my original iPod at a reasonable cost? I am, after all, still listening to the same music. Did I bother to mention that I can't even purchase an operating system for my Apple G5 computer? My new iPod doesn't work unless I upgrade my software. Do I now need to throw away my computer and my old iPods since Apple now only supports Intel chip sets? Apple engineers itself this way. It is not an accident. It is a designed throw-away company to generate more sales year on year.

Mercedes Benz is held up for its excellence in engineering. I bought into the perception and my wife and I purchased an A class for her and an ML 4WD for myself in 2001. Both cars were an engineering nightmare. Without going into too much detail, the pièce de résistance was the drainage pipe from the sunroof in the ML. It was attached with only a barb on the end of a plastic hose to keep it in place. This worked itself free so water would pour onto the car's computer system and fuse box. I'm not alone; this is a world-wide problem that Mercedes didn't address.[98] The repair would have cost Mercedes Benz less than 1c for a cable tie on the pipe. Such a basic engineering mistake must have been intentional. Mercedes designed its car to fail so you needed to purchase a new one.

Remember my Massey Ferguson tractor from Chapter 4 the same type that Edmond Hillary drove around Antarctica? When something breaks on this 55 year old machine, I simply drive over to the local tractor shop and pick up the part. Almost always they have the broken part in stock. Worst case, it is available the next day. And they are cheap. Sometimes major components are less than $10.00. Spare parts? What a great concept.

When the key to my Mercedes Benz failed to work because of the water leakage into the computer, the part had to come from Germany! You read it correctly; a key had to come to Australia from Germany. And it cost over $1000 and it took six weeks to arrive. My car had to sit in the ski fields, 5 hours from my home, waiting for nothing more than a computer chip. And after all this, did Mercedes Benz fix the hose? No, it fell off again. This time I fixed it myself. No one from Mercedes would tell me how to do it - the throw-away engineering company – Mercedes Benz.

How many of us have purchased new plasma high definition televisions screens? One of my HDMI ports has stopped working. I can't get this fixed without spending a small fortune and the set is only 2 years old. These televisions are going to all fail on mass over the next decade. They will not be able to be fixed. Throw them away and start again is the design plan.

How many small plastic parts break on children's toys and the entire toy has to be thrown away? With *Make it Maintain it*, everything built in the future has to come with parts lists, manufacturing diagrams, servicing instructions and a stockpile of spare parts. Moulds for all plastic components need to be supplied with imported product shipments. Nothing can be sold without the ability for it to be maintained.

Why can't we have a standard low voltage power supply, with a common connection? Why each time you purchase a phone, laptop, camera, iPod or GameBoy do you also receive a new power supply? Like mains power, why can't we have a standard connection? I know the answer, it's because electronics companies view the connection as a marketing advantage. In *Make it Maintain it*, we would have established standards for power supplies and a universal connection plug.

Open systems and standards made the PC industry universal, why can't such concepts be transferred to the electronics, automotive and white goods industries?

Think of how much less landfill we would need for trash and how much pollution would we reduce?

Governments could enact legislation. Something they do well. They could give industry 3 years to get used to the concept. Engineering for maintenance rather than obsolesce. Established standards will be established for interconnection. This alone would create thousands of jobs.

Also, this concept doesn't need the world's agreement to work.

*Make it Maintain it* would be good for the local economies as well. Maintenance jobs would be created and new maintenance based design and repair concepts would be required. Greeters at Wal-Mart could get new jobs fixing stuff - if they wanted to.

It used to be this way. Even when technology continuously changed in the 1980s, IBM reused its mainframe technology again and again. It was called an upgrade. Now it's a replacement.

Think again about how much we throw away? A cut of 50% in our carbon footprint would be simple if we upgraded rather than replaced.

My other sister Ann lives in Phoenix, Arizona. And while the growth of her metropolitan area has slowed somewhat over the last few years, the region's population is still around 4.3 million people - all living in the hot desert, requiring electricity for air conditioning and even heating in the winter. And did I mention water? All pumped in.

For me, living in Australia for the past 20 years, I got used to the fact that every home has a clothes line *'out the back.'* The thought of using an electric heated clothes dryer was totally foreign in Australia when I arrived, but has crept into more and more homes over the years. Ann's home in Phoenix has no clothes line. She lives in the desert and clothes would dry in a matter of minutes. Yet they all go into the clothes drier. For me, clothes smell fresh and they don't shrink from being hung on the line. After asking her about a clothes line, I'm told that they are not allowed in her community. I later learn that hanging clothes outside is somehow a sign of poverty and council by-laws in her area prohibited clothes to be hung out to dry.

If we eliminated clothes driers and excess toy packaging just think how low our carbon footprint would be?

Contrast the two models:

A tax is placed on carbon. This is watered down for developing countries. We ship our coal to China and India and the pollution continues unabated. Electricity prices at home soar. Solar systems fail to deliver any cost savings. We are all in debt for thousands of dollars. Cheap products continue to flood in from China and India. The world is polluted.

Or.

Maintenance standards are established. The price of electricity doesn't increase. Some products may initially go up in price, but market forces correct this because the cost to maintain is not actually a heavy burden. More jobs are created in the West. Pollution levels drop. The world is a cleaner place.

Look at Pittsburgh today. It's a clean and vibrant city. The woodland behind my father's house has transformed. It now has wild turkeys and deer frequently drop by to eat his flowers. It wouldn't surprise me if the black bear returns. We can clean the world up and continue to prosper. All we need is some will.

To grow this will from the grass roots is much better for society than resorting to *hyper-inflated marketing*. The future of the planet is in your hands.

# FIGURE INDEX

# INDEX

# REFERENCES

[1] Australian Bureau of Metrorology. *Climate Data Online.*
http://www.bom.gov.au/climate/data/ [Accessed: 2010].

[2] Lester, B. (2006). Australia's drought may stay for keeps. *Cosmos.* 15 December 2006
http://www.cosmosmagazine.com/node/927 [Accessed: 14 Nov 2007].

[3] Sydney Water. (2010). *Desalination.*
http://www.sydneywater.com.au/Water4Life/desalination/ [Accessed: 3 December 2010].

[4] Bita, N. (2010). Water charges are set to spiral in desalination squeeze *The Australian*, 23
Jan 2010. http://www.theaustralian.com.au/news/nation/water-charges-are-set-to-spiral-in-desalination-squeeze/story-e6frg6nf-1225822705341 [Accessed: 28 Feb 2011].

[5] Sydney Catchment Authority. *Dam Levels.* http://www.sca.nsw.gov.au/ [Accessed: 3 Jan
2011].

[6] Om, J. (2010). Record rain not enough to end drought. *ABC News.*
http://www.abc.net.au/news/stories/2010/10/01/3027192.htm [Accessed: 5 December
2010].

[7] IPCC (2007a). Summary for Policymakers. *In: Climate Change 2007: The Physical Science
Basis. Contribution of Working Group I to the Fourth Assessment Report of the
Intergovernmental Panel on Climate Change [Solomon, S., D. Qin, M. Manning, Z. Chen, M.
Marquis, K.B. Averyt, M.Tignor and H.L. Miller (eds.)]. Cambridge University Press,
Cambridge, United Kingdom and New York, NY, USA.* .
http://www.ipcc.ch/pdf/assessment-report/ar4/wg1/ar4-wg1-spm.pdf [Accessed: 15 Jun
2008].

[8] Solomon, S., Qin, D., M. Manning, R.B. Alley, T. Berntsen, N.L. Bindoff, Z. Chen, A.
Chidthaisong, J.M. Gregory, G.C. Hegerl, M. Heimann, B. Hewitson, B.J. Hoskins, F. Joos, J.
Jouzel, V. Kattsov, U. Lohmann, T. Matsuno, M. Molina, N. Nicholls, J. Overpeck, G. Raga,
V. Ramaswamy, J. Ren, M. Rusticucci, R. Somerville, T.F. Stocker, P. Whetton, Wood, R. A.
& D. Wratt. (2007). *Technical Summary. Climate Change 2007: The Physical Science Basis.
(AR4-TS)* Intergovernmental Panel on Climate Change, World Meteorological Organization
(WMO) and the United Nations Environment Programme (UNEP). United Kingdom.
Cambridge University Press.
http://www.ipcc.ch/publications_and_data/publications_and_data_reports.shtml.

[9] Holper, P. & Torok, S. (2008). *Climate Change - What you can do about it*, CSIRO
Publishing.

[10] Vaughan, D. G., Marshall, G. J., Connolley, W. M., King, J. C. & Mulvaney, R. (2001). Devil
in the Detail. *Science*, 293, 1777-1779.
http://www.sciencemag.org/content/293/5536/1777.citation [Accessed: 1 July 2008].

[11] Vaughan, D. G., Marshall, G. J., Connolley, W. M., Parkinson, C., Mulvaney, R., Hodgson,
D. A., King, J. C., Pudsey, C. J. & Turner, J. (2003). Recent Rapid Regional Climate Warming
on the Antarctic Peninsula. *Climatic Change*, 60, 243-274.
http://www.antarctica.ac.uk/bas_research/science/climate/antarctic_peninsula.php
[Accessed: 1 Aug 08].

[12] Turner, J., Colwell, S. R., Marshall, G. J., Lachlan-Cope, T. A., Carleton, A. M., Jones, P. D.,
Lagun, V., Reid, P. A. & Iagovkina, S. (2005). Antarctic climate change during the last 50
years. *International Journal of Climatology*, 25, 279-294.

www.scar.org/researchgroups/physicalscience/reader_turneretal.pdf [Accessed: 20 July 2009].

[13] Vaughan, D. G. (2006). Recent trends in melting conditions on the Antarctic Peninsula and their implications for ice-sheet mass balance. *Arctic, Antarctic and Alpine Research,* 38, 147-152. http://www.jstor.org/pss/4095837 [Accessed: 20 July 2009].

[14] BAS-READER. (2008). *Antarctic meteorology; Surface tempertures*. British Antarctic Survey, Natural Environment Research Council. http://www.antarctica.ac.uk/met/gjma/ [Accessed: 12 Aug 08].

[15] GSOD. (2008). *NNDC Climate Data Online, Global Summary of the Day (GSOD)*. National Environmental Satelitte, Data , and Information Service (NESDIS). http://www7.ncdc.noaa.gov/CDO/cdoselect.cmd?datasetabbv=GSOD&countryabbv=&georegionabbv= [Accessed: 12 Aug 08].

[16] BAS-READER. (2008). *Antarctic meteorology; Surface tempertures*. British Antarctic Survey, Natural Environment Research Council. http://www.antarctica.ac.uk/met/gjma/ [Accessed: 12 Aug 08].

[17] Vaughan, D. G., Marshall, G. J., Connolley, W. M., King, J. C. & Mulvaney, R. (2001). Devil in the Detail. *Science,* 293, 1777-1779. http://www.sciencemag.org/content/293/5536/1777.citation [Accessed: 1 July 2008].

[18] (2006b). Antarctic will melt as carbon dioxide levels rise: scientists. *ABC News (Australia)*, 13 July 2006. http://www.abc.net.au/news/newsitems/200607/s1685513.htm [Accessed: 5 December 2010].

[19] (2009). Greenland and Antarctic Ice Sheet Melting, Rate Unknown. *ScienceDaily*. 25 February 2009 http://www.sciencedaily.com/releases/2009/02/090216131158.htm [Accessed: 30 November 2010].

[20] Munro, M. (2009). Antarctic ice melt could shift Earth's rotation: Study. *Canwest News Service*, 5 February 2009. http://www2.canada.com/topics/technology/science/story.html?id=1261393 [Accessed:

[21] Australian Bureau of Meteorology. *Climate Data Online*. http://www.bom.gov.au/climate/data/ [Accessed: 2010].

[22] Bernstein, L. & all, e. (2007). *Climate Change 2007. Synthesis Report - Summary for Policymakers*. Intergovernmental Panel on Climate Change, World Meteorological Organization (WMO) and the United Nations Environment Programme (UNEP). United Kingdom. Cambridge University Press. http://www.ipcc.ch/publications_and_data/publications_ipcc_fourth_assessment_report_synthesis_report.htm. [Accessed 2009]

[23] National Climatic Data Centre. (2010). *Global Historical Climatology Network - Daily*. NOAA Satellite and Information Service. http://www.ncdc.noaa.gov/oa/climate/ghcn-daily/ [Accessed: 15 December 2010].

[24] CIA. (2010). *The World Factbook*. Central Intelligence Agency. www.cia.gov/library/publications/the-world-factbook/geos/xx.html [Accessed: 19 March 2010].

[25] BAS-READER. (2008). *Antarctic meteorology; Surface tempertures*. British Antarctic Survey, Natural Environment Research Council. http://www.antarctica.ac.uk/met/gjma/ [Accessed: 12 Aug 08].

[26] Ibid.

[27] Vaughan, D. G., Marshall, G. J., Connolley, W. M., King, J. C. & Mulvaney, R. (2001). Devil in the Detail. *Science,* 293, 1777-1779. http://www.sciencemag.org/content/293/5536/1777.citation [Accessed: 1 July 2008].

[28] Graham, S. (2002). Climate warming causes collapse of antarctic ice shelf *Scientific American*. http://www.scientificamerican.com/article.cfm?id=climate-warming-causes-co [Accessed: 15 Dec 2010].

[29] British Antarctic Survey (2006). First Direct Evidence that human activity is linked to Antarctic ice shelf collapse. *Journal of Climate*. http://www.sciencedaily.com/releases/2006/10/061016105739.htm [Accessed: 19 Dec 2010].

[30] Williams, J. (2004). Ice shelf breakup challenges researchers. *USA Today*, 26 May 2004. http://www.usatoday.com/weather/resources/coldscience/2004-05-26-peninsula-conf_x.htm [Accessed: 30 Oct 2010].

[31] NASA. (2010). *World Book at NASA - Global Warming*. http://www.nasa.gov/worldbook/global_warming_worldbook.html [Accessed: 19 Dec 2010].

[32] Glasser, N. & Scambos, T. (2008). Antarctic ice shelf collapse blamed on more than climate change. *Journal of Glaciology*. http://www.sciencedaily.com/releases/2008/02/080210100441.htm [Accessed: 19 Dec 2010].

[33] Belville, J. H. (1850). *The Thermometer; containing its history and use as a meteorological instrument,* London, http://books.google.com.au/books.

[34] UK Met Office. (2010). *Climate Change*. http://www.metoffice.gov.uk/climatechange/science/explained/explained5.html [Accessed: 5 Dec 2010].

[35] Oxburgh, R., Davies, H., Emanuel, K., Graumlich, L., Hand, D., Huppert, H. & Kelly, M. (2010). *Report of the International Panel set up by the University of East Anglia to examine the research of the Climatic Research Unit*. University of East Anglia.London. http://en.wikipedia.org/wiki/Climatic_Research_Unit_email_controversy [Accessed: 12 Dec 2010].

[36] Australian Bureau of Metrorology. *Climate Data Online*. http://www.bom.gov.au/climate/data/ [Accessed: 2010].

[37] Black, R. (2009). *Why did Compenhagen fail to deliver a climate deal?* BBC News. http://news.bbc.co.uk/2/hi/8426835.stm [Accessed: 3 December 2010].

[38] (2000). *Climate Change and the Pacific Islands.* Ministerial Conference on Environment and Develoment in Asia and the Pacific 2000, 31 August - 5 September 2000 2000 Kitakyushu, Japan. http://www.unescap.org/mced2000/pacific/background/climate.htm [Accessed: 7 December 2010].

[39] Associated Press. (2010). Study: Coral atolls hold on despite sea-level rise. *Samoa News.com*, 3 June 2010. http://www.samoanews.com/viewstory.php?storyid=15707&edition=1275559200 [Accessed: 12 Dec 2010].

[40] Meares, R. (2009). Global warming may bring tsunami and quakes: scientists. *Reuters*. http://in.reuters.com/article/idINTRE58F62I20090916 [Accessed: 30 Oct 2010].

[41] McCarragher, B. *New Orleans Hurricane History*. MIT. http://web.mit.edu/12.000/www/m2010/teams/neworleans1/hurricane%20history.htm [Accessed: 5 December 2010].

[42] National Hurricane Centre. (2010). *Hurricane History*. National Weather Service. http://www.nhc.noaa.gov/pastall.shtml [Accessed: 5 December 2010].

[43] Australian Bureau of Metrorology. (2010b). *Tropical Cyclones*. BOM. http://www.bom.gov.au/cyclone/index.shtml [Accessed: 5 December 2010].

[44] Love, G. B. D. (2006). *Statement on Tropical Cyclones and Climate Change*. WMO/CAS Tropical Meteorology Research Program. http://www.com.gov.au/info/CAS-statement.pdf.

[45] Love, G. D. & McBride, J. D. (2006). *Media Release - Tropical cyclones and climate change*. BOM. http://www.bom.gov.au/announcements/media_releases/ho/20060220.shtml.

[46] World Meteorological Organization. (2010). *World Data Centre for Greenhouse Gases*. http://gaw.kishou.go.jp/wdcgg/ [Accessed: 9 Dec 2010].

[47] IPCC. (2007b). *What is the Greenhouse Effect*. http://www.ipcc.unibe.ch/publications/wg1-ar4/faq/wg1_faq-1.3.html [Accessed: 12 Dec 2010].

[48] Holper, P. & Torok, S. (2008). *Climate Change - What you can do about it*, CSIRO Publishing.

[49] Vardanyan, M., Trotta R & Silk, J. (2011). Applications of Bayesian model averaging to the curvature and size of the Universe. *Cosmology and Extragalactic Astrophysics*. http://arxiv.org/abs/1101.5476 [Accessed: 28 Feb 2011].

[50] Stott, P. A., Jones, G., S., & Mitchell, J. F. B. (2003). Do Models Underestimate the Solar Contribution to Recent Climate Change. *Journal of Climate*, 16. [Accessed:

[51] Ravilious, K. (2007). Mars Melt Hints at Solar, Not Human, Cause for Warming, Scientist Says. *National Geographic News*. http://news.nationalgeographic.com/news/2007/02/070228-mars-warming.html [Accessed: 30 Oct 2010].

[52] WPXI. (2007). *Report Released on Effects of Global Warming in Pennsylvania*. WPXI. http://www.wpxi.com/station/13665170/detail.htm [Accessed: 11 December 2010].

[53] Hopey, D. (2006). Study says Pittsburgh to be one hot town. *Pittsburgh Post Gazette*, 6 October 2006. http://www.post-gazette.com/pg/06279/727848-113.stm [Accessed: 15 Nov 2010].

[54] Potter, N. (2010). Winter Weather Update: Hard Freeze from Canada to Gulf of Mexico. *ABC News*. http://abcnews.go.com/Travel/winter-weather-cold-ice-snow-continue-georgia-pennsylvania/story?id=9513139 [Accessed: 3 Dec 2010].

[55] GSOD. (2008). *NNDC Climate Data Online, Global Summary of the Day (GSOD)*. National Environmental Satelitte, Data , and Information Service (NESDIS). http://www7.ncdc.noaa.gov/CDO/cdoselect.cmd?datasetabbv=GSOD&countryabbv=&georegionabbv= [Accessed: 12 Aug 08].

[56] Doughton, S. (2007). Kilimanjaro not a victim of climate change, UW scientist says. *The Seattle Times*. http://seattletimes.nwsource.com/html/localnews/2003744089_kilimanjaro12m.html [Accessed: 12 Dec 2010].

[57] Ibid. [Accessed: 12 Dec 2010].

[58] (2006a). www.cuso4.org/photos/eu20060609/20060609-fr-chamonix-09m.jpg [Accessed: 4 April 2010].

[59] Ramming, A., Jonas, T., Zimmermann, N. E. & Rixen, C. (2010). Changes in alpine plant growth under future climate conditions. *Biogeosciences*, 2013-2024. http://www.biogeosciences.net/7/2013/2010/bg-7-2013-2010.html [Accessed: 3 Mar 2011].

[60] Watts, A. (2009). The Harris Poll. *Financial Times*, 22 Oct 2009. http://wattsupwiththat.com/2009/10/22/harris-poll-europeans-tend-to-care-more-strongly-about-climate-change-than-americans/ [Accessed: 3 Mar 2011].

[61] CBC News. (2010a). Polar bears invade Nunavut areas. *CBC News*, 9 Dec 2010. http://www.cbc.ca/canada/north/story/2010/12/09/nunavut-polar-bear-problems.html [Accessed: 21 Dec 2010].

[62] Young, E. (2002). Climate change threatens polar bears. *NewScientist*. 15 May 2002 http://www.newscientist.com/article/dn2285-climate-change-threatens-polar-bears.html [Accessed: 21 Dec 2010].

[63] Roach, J. (2007). Most polar bears gone by 2050, studies say. *National Geographic*. 10 Sep 2010 http://news.nationalgeographic.com/news/2007/09/070910-polar-bears.html [Accessed: 21 Dec 2010].

[64] Kattsov, V., M., & Källén, E. (2004). Future Climate Change: Modeling and Scenarios for the Arctic. *Arctic Climate Impact Assessment (ACIA),* Chapter 4. http://www.acia.uaf.edu/PDFs/ACIA_Science_Chapters_Final/ACIA_Ch04_Final.pdf [Accessed: 21 Dec 2010].

[65] Folger, T. (2010). Changing Greenland. *National Geographic*. June 2010 http://ngm.nationalgeographic.com/2010/06/viking-weather/folger-text/1 [Accessed: 15 Dec 2010].

[66] Traufetter, G. (2006). Global Warming a Boom for Greenland Farmers. *Der Spiegel*. 30 Aug 2006 http://www.spiegel.de/international/spiegel/0,1518,434356,00.html [Accessed: 15 Dec 2010].

[67] Folger, T. (2010). Changing Greenland. *National Geographic*. June 2010 http://ngm.nationalgeographic.com/2010/06/viking-weather/folger-text/1 [Accessed: 15 Dec 2010].

[68] Potter, N. (2010). Winter Weather Update: Hard Freeze from Canada to Gulf of Mexico. *ABC News*. http://abcnews.go.com/Travel/winter-weather-cold-ice-snow-continue-georgia-pennsylvania/story?id=9513139 [Accessed: 3 Dec 2010].

[69] CBC News. (2010b). North Pole rainfall 'bizarre' climatologist. 29 April 2010. http://www.cbc.ca/canada/north/story/2010/04/29/north-pole-rainfall.html [Accessed: 23 Dec 2010].

[70] NASA. (2009a). *NASA satellite reveals dramatic arctic ice thinning.* http://www.nasa.gov/topics/earth/features/icesat-20090707r.html [Accessed: 21 Dec 2010].

[71] Kwok, R. & Rothrock, D. A. (2009). Decline in Arctic sea ice thickness from submarine and ICESat records 1958-2008. *Geophysical Research Letters,* 36. http://rkwok.jpl.nasa.gov/publications/Kwok.2009.GRL.pdf [Accessed: 21 Dec 2010].

[72] (2010a). *Operation Wunderland*. http://www.allworldwars.com/Operation-Wunderland-1942.html [Accessed: 28 Dec 2010].

[73] NASA. (2009b). *What's holding Antarctic sea ice back from melting?* http://earthobservatory.nasa.gov/Newsroom/view.php?id=40042 [Accessed: 21 Dec 2010].

[74] CBC News. (2010a). Polar bears invade Nunavut areas. *CBC News*, 9 Dec 2010. http://www.cbc.ca/canada/north/story/2010/12/09/nunavut-polar-bear-problems.html [Accessed: 21 Dec 2010].

[75] CSIRO. (2010). *State of the Climate*. CSIRO and Bureau of Meteorology. http://www.csiro.au/news/State-of-the-Climate.html. [Accessed 10 Nov 2010]

[76] (2010b). *What is Coral Bleaching*. Australian Government, Great Barrier Reef Marine Park Authority. http://www.gbrmpa.gov.au/corp_site/key_issues/climate_change/climate_change_and_the_great_barrier_reef/what_is_coral_bleaching [Accessed: 5 Feb 2011].

[77] Berkelmans, R., De'ath, G., Kininmonth, S. & Skirving, W. J. (2004). A comparison of the 1998 and 2002 coral bleaching events on the Great Barrier Reef: spatial correlation, patterns, and predictions. *Coral Reefs, 23*, 74-83. http://imars.usf.edu/~cmoses/PDF_Library/Berkelmans%20et%20al%202004.pdf [Accessed: 5 Feb 2011].

[78] Hoegh-Guldberg (1999). Climate change, coral bleaching and the future of the world's coral reefs. *Freshwater,* Mar, 839-866. www.reef.edu.au/climate/Hoegh-Guldberg%201999.pdf [Accessed: 5 Feb 2011].

[79] Gleason, D. F. & Wellington, G. M. (1993). Ultraviolet radiation and coral bleaching. *Nature, 365,* 836-8. http://www.nature.com/nature/journal/v365/n6449/abs/365836a0.html [Accessed: 12 Jan 2011].

[80] (2010b). *What is Coral Bleaching.* Australian Government, Great Barrier Reef Marine Park Authority. http://www.gbrmpa.gov.au/corp_site/key_issues/climate_change/climate_change_and_t he_great_barrier_reef/what_is_coral_bleaching [Accessed: 5 Feb 2011].

[81] Hoegh-Guldberg (1999). Climate change, coral bleaching and the future of the world's coral reefs. *Freshwater,* Mar, 839-866. www.reef.edu.au/climate/Hoegh-Guldberg%201999.pdf [Accessed: 5 Feb 2011].

[82] Downing, N., Buckley, R., Stobart, B., LeClair, L. & Teleki, K. (2004). Reef fish diversity at Aldabra Atoll, Seychelles, during the five years following the 1998 coral bleaching event. *Philosophical Transactions of the Royal Society, Mathematical, Physical and Engineering Sciences, 363,* 257-261. [Accessed: 5 Feb 2011].

[83] Minkel, J. R. (2007). Confirmed: The US Census Bureau gave up names of Japanese-Americans in WWII. *Scientific American.* 30 March 2007 http://www.scientificamerican.com/article.cfm?id=confirmed-the-us-census-b&sc=I100322 [Accessed: 24 Dec 2010].

[84] Bender, B. (2003). Democracy might be impossible, US was told. *Boston Globe,* 14 Aug 2003. http://www.commondreams.org/headlines03/0814-06.htm [Accessed: 24 Dec 2010].

[85] Watson, C. (2010). Fears of stranger danger is restricting children's freedom. *Adelaide Advertiser.* http://www.adelaidenow.com.au/news/south-australia/fears-of-stranger-danger-is-restricting-childrens-freedom/story-e6frea83-1225935127717 [Accessed: 24 Dec 2010].

[86] Gordon, J. (2010). Canada urged to ban asbestos mining, exports *Reuters,* 30 Jun 2010. http://www.reuters.com/article/2010/06/30/us-asbestos-idUSTRE65T4UM20100630 [Accessed: 18 Feb 2010].

[87] Cook, A. (2010). BoM and CSIRO remind Australian Ski Industry of the bad news. *MountainWatch,* 22 Mar 2010. http://www.mountainwatch.com/Features/7347130/BoM-and-CSIRO-remind-Australian-Ski-Industry-of-the-bad-news [Accessed: 4 Jan 2011].

[88] Australian Bureau of Metrorology. (2010a). *Queensland Flood History.* http://www.bom.gov.au/hydro/flood/qld/fld_history/index.shtml [Accessed: 5 Feb 2011].

[89] Gibbs, W. J., Shields, A. J., Neal, A. B. & G , H. (1974). *Brisbane Floods January 1974.* Meteorology, B. O., Canberra. Australian Government Publishing Service. http://www.bom.gov.au/hydro/flood/qld/fld_reports/brisbane_jan1974.pdf. [Accessed 8 Feb 2011]

[90] Andersen, B. (2011). Wivenhoe put to ultimate test. *ABC News,* 11 Jan 2011. http://www.abc.net.au/news/stories/2011/01/11/3110758.htm [Accessed: 5 Feb 2011].

[91] Thomas, H. (2011). Engineer's emails reveal Wivenhoe Dam releases too little, too late. *The Australian*, 21 Jan 2011. http://www.theaustralian.com.au/in-depth/queensland-floods/engineers-emails-reveal-wivenhoe-dam-releases-too-little-too-late/story-fn7iwx3v-1225991990957 [Accessed: 22 Jan 2011].

[92] Gearing, A. (2007). *Bring us a monsoon,* The Courier-Mail, Brisbane, 10-11 Feb 2007.

[93] Seqwater. (2010). *Seqwater 2010-11 to 2014-15 Strategic Plan.* http://www.seqwater.com.au/public/sites/default/files/userfiles/file/pdfs/Seqwater2010-15%20Strategic%20Plan%20final.pdf. [Accessed 8 Feb 2011]

[94] Garnaut, R. (2008). *The Garnaut Climate Change Review.* http://www.garnautreview.org.au/index.htm [Accessed: 27 Dec 2010].

[95] Bennett, B., Repacholi, M. & Carr, Z. (2006). *Health Effects of the Chernobyl Accident and Special Health Care Programme.* World Health Organization, Geneva. http://www.who.int/ionizing_radiation/chernobyl/WHO%20Report%20on%20Chernobyl%20Health%20Effects%20July%2006.pdf. [Accessed 12 Dec 2010]

[96] AFP. (2011). IAEA warned Japan over nuclear quake risk: WikiLeaks. 17 March 2011. http://www.google.com/hostednews/afp/article/ALeqM5i60CuTNdU7t3Ok_vSQCk_jsDUKpg?docId=CNG.9f3603fc4dd3586217c60241e5e5eb79.281 [Accessed: 17 Mar 2011].

[97] Bhavnagri, K. (2010). *Home energy consumption versus solar PV generation.* http://www.solarchoice.net.au/blog/home-energy-consumption-versus-solar-pv-generation.html [Accessed: 5 Dec 2010].

[98] Mercedes Benz Owners Community. (2010). http://www.mercedes-benz-usa.com/ml_class.php [Accessed: 12 Feb 2010].

www.ingramcontent.com/pod-product-compliance
Lightning Source LLC
Chambersburg PA
CBHW072303210326
41519CB00057B/2607